JAMES COOK
MARITIME SCIENTIST

CAPTAIN

Jam Cook

F.R.S.

CAPTAIN COOK. Taken from George Young's *History of Whitby,* which in its turn was taken from the Dance portrait, commissioned by Sir Joseph Banks.

JAMES COOK
MARITIME SCIENTIST

TOM AND CORDELIA STAMP

'A man in nautical knowledge inferior to none. In
Zeal, Prudence, and Energy, superior to most.'

Robert Campion, 1827

𝕮𝖆𝖊𝖉𝖒𝖔𝖓 𝖔𝖋 𝖂𝖍𝖎𝖙𝖇𝖞

Whitby, Yorkshire, England.

By the same authors:

William Scoresby Arctic Scientist
Caedmon of Whitby Press   1976

ISBN   0 905355 04 0
© Caedmon of Whitby Press   1978

Published by: Caedmon of Whitby Press
Harold Villa, Whitby, Yorkshire
YO21 3AP England.

Printed by: The Scolar Press Limited,
Ilkley, West Yorkshire.

FOR FRANCISCA, WITH LOVE

ACKNOWLEDGEMENTS

Our thanks are due to many people who have helped in the making of this book and we should especially like to thank:

Michael Dawson and the Yorkshire Arts Association for financial and moral support.

Mr A. K. Cumbor of Great Ayton, and the Cook Schoolroom Trust for the pictures of the cottage being dismantled.

Dr. A. M. Luyendijk-Elshout of Leiden for helping in the Bachstrom search.

Fred Fletcher of Hull for detailed and exact extracts from early Quaker minutes; the librarian of Friends House, London, for prompt replies to many queries.

Mr J. G. Graham and Mr G. E. Gunner of the Whitby Literary and Philosophical Society, whose encouragement and extremely practical assistance in the correction of the several variations of the original manuscript were quite invaluable.

Professor Sir Alister Hardy for many kind actions, but chiefly for the tremendous encouragement which his concern for our venture gave us.

Mr W. Leng and Mr I. F. Brocklesby for help with the photographs.

Mrs Helga Perkins and her son Gottfried for translating the Bachstrom Biography.

The manager and staff of the Office Equipment Bureau of Scarborough, who ably met the numerous requests of a temperamental, arthritic typist.

Mr C. Thornton of the Dorman Museum in Middlesbrough, for his liberal sharing of his extensive Cook knowledge.

Miss N. Vickers and her staff at the Whitby Library, who have never let us down yet, and on request have produced the most obscure and remote books as if by magic.

CONTENTS

ILLUSTRATIONS

INTRODUCTION

This book has been written from the viewpoint of dwellers on the North East coast of Britain, and especially of Whitby. We two are proud of our association with the old town.

It is a remarkable fact that the small seaport of Whitby, — for centuries relatively isolated — does nevertheless hold a recognised place in the religious and maritime history of our island peoples.

To have been the scene of the Synod of Whitby and the site of Hilda's Abbey and the monastic foundation that followed; to have nurtured Caedmon, the first English poet; to have fostered the genius of Cook, giving him his marriage certificate to the sea and building his ships; to have fathered the fine abilities of the Scoresbys and bred the hardy, adventurous spirits that went with them to the Greenland seas, — these are no small claims to fame and to our affection.

It was from this ancient seaport that Cook sailed and learnt his seaman's skill. Here were spent the all important years that turned him from an inland youth to a man of the sea. The process began at Staithes among the fishing folk and small craft but it was here in Whitby that James Cook was irrevocably wed to the sea. And if that were not claim to fame enough, Whitby built his ships. Aye, all four of them which sailed in his great voyages were built here.

The choice of Whitby ships was no mere coincidence, but a deliberate act of Cook, based on his experience and sound judgement, and, as he had foreseen, they turned out to be the very ships ideally suited to their task. So, with Whitby ships, Cook blazed a trail round the world; opened up the unknown southern ocean; discovered and charted the east coast of Australia and the islands of New Zealand and the Pacific; confirmed Bering's discovery of his Strait; set limits to Antarctica, and much besides.

Can anyone wonder that we here in Whitby are proud of our connection with James Cook?

A major interest in Cook's time was geographical exploration. The natural sciences had the whole world laid before them, and motivated and inspired by the colonizing and empire building spirit of the period, men moved out to find whatever the most remote regions could offer in the way of new plants and animal and mineral wealth. To achieve their ambitions, long sea voyaging was essential, and an absolute prerequisite to success in such voyages was the control of scurvy and the maintenance of the health of seamen. In all these fields Cook made outstanding contributions.

His thorough testing of Lind's theories and his proof of their efficacy in practice constitute a vital addition to knowledge. That the achievement is solely his, we see by the incident in Chapter 9 when the two ships were separated for a time. When they met again it was found that scurvy had broken out in the Adventure, while Cook's personal supervision of his dietary and sanitary rules had kept the Resolution clear of it.

Another important advance in the last half of the eighteenth century was the accurate determination of longitude. Here again Cook, though not the innovator, was the first and most able practitioner of the new methods and did most to demonstrate their utility and accuracy. His mastery in the first voyage of the lunar method of determining longitude and his convincing demonstration during the second that the Harrison-Kendall chronometer had solved the problem, enabled him to pin-point his discoveries and chart them with great accuracy. For though others had tested chronometers, none had subjected them to such stringent and varied tests over such long periods and in so many different climatic conditions as Captain Cook.

The natural scientists of that time had whole continents to explore and the romance and adventure of the quest reached its grandest proportions in the life and work of James Cook, who sailed into the largely mysterious southern ocean, unknown and uncharted, vague and remote, the subject of the wildest flights of geographical imagination, and at the end came back into the western world with maps and charts, with soundings and descriptions, knowledge and certainty. His has remained an unparalleled life of discovery and exploration beyond any before or since.

There was one other outstanding man on the Endeavour. It was no mere adventurous whim that took the young Joseph Banks on the memorable first voyage and started him on his fruitful career as botanist extraordinary and scientific adviser to the State. Banks became President of the Royal Society and a personal friend of George III, and he exercised a profound and beneficial influence on the course of British science, being, in effect, the king's adviser on all matters relating to science. His expert knowledge of estate management was an additional bond of friendship with 'Farmer George'.

There is little doubt that Captain Cook and Joseph Banks understood one another perfectly. The difficulties which arose over the arrangements for the proposed second voyage never seriously impaired their high regard for each other. We must remember that it was Banks who commissioned the Dance portrait of Cook. Cook valued and esteemed the younger man's enthusiastic adventurousness, his earnest desire to promote science and exploration; his amiable and outreaching personality and his ability to recognise the scientific potential in the gifts of people of different walks of life, − in short, the qualities that made Banks an outstanding Georgian statesman of science.

We may well suppose Banks to have had a special interest in Whitby. Had not the *Endeavour* been built there, and its great commander, his valued friend, formed strong ties from his youth with the old town? Did not his aristocratic and accomplished friend, C. J. Phipps, have an estate nearby where he had stayed while Omai was with him? Small wonder that in his later life Banks should befriend the Scoresbys, who also hailed and sailed from Whitby.

It was Banks who supported the King's appointment of William Herschel as Royal Astronomer. There is a striking likeness in the intellectual achievements of James Cook (1728–1779) and William Herschel (1738–1822). Both were men of a strong practical turn of mind, yet both possessed fine brains and breadth of vision. Both acquired the intellectual equipment for their life's work by a commendable effort of self-help in education, and by mastering observational and manipulative skills. One set out to explore the oceans and the lands that lay beyond, and the other the universe at large. In our own day the two types of scientific activity are combined in the efforts of the space explorers of the solar system.

Apart from the possibility of early Quaker influence at Ayton, there can be no doubt that Cook's coming to the Walker household at Whitby helped to shape his character along Quaker lines. The clearer our picture of Cook's personality and life style becomes, the more of Quaker qualities and attributes do we find there; the long silences; the periods of inner reflectiveness; the abstention from swearing, however provoked; the dislike of ritual and priests in his ship; the careful, sparing use of words in speech; the recognition of the human worth of all men, sailors and savages alike; these are all marks of the true Quaker. James Cook, had he joined the Society of Friends, would have made an outstanding member.

Many biographers have stressed the enigmatic qualities of Cook the man. How difficult, they say, to penetrate that mask of personal reserve. To a certain extent they are right for, indeed, men of character and intellect are enigmatic to those who seek to understand from the viewpoint of every day life and experience. Yet it is from such a viewpoint that the human qualities of any man, however gifted, should most surely emerge. This is certainly true of Cook, for we find the restraint is merely the natural reserve of a man who did not reveal himself easily; the caution of one who habitually weighed all contingencies and eventualities; who thought more deeply, saw more clearly and acted more resolutely than any around him.

We have tried to give a portrait of James Cook himself and not to be side tracked by other personalities. We have said very little about people who were close to Cook, such as Hugh Palliser or Charles Clerke, which has been difficult, for those whose lives touch ours often have a great influence upon our thoughts and actions; but, for one thing we wanted this book to be short and easily read, and for another, it might have obscured the clear picture of the man which becomes apparent from a study of his writings.

With the exception of proper names, which appear as Cook wrote them, we have used modern spelling as we felt the older spelling only serves to put a barrier between the author and the reader.

We have drawn upon three main sources for our work. Firstly, upon a Whitby historian, the Reverend George Young, whose *Life and Work of Captain James Cook* was written at a time when living memory could recall the man and his times. Young himself was a most interesting figure. He was the minister of Cliff Lane Chapel, in Whitby, from 1806 until his death in 1848. He talked to many people who had actually known Cook and was thus able to gain first hand information about his early, formative years. As well as being a careful historian and a scholar, Young was also a very human man, and it is recorded that his people looked upon him as a friend as well as a minister. In 1817 he published an exhaustive *History of Whitby*, the writing of which was preceded by years of research, careful study and verbal questioning of local people. He brought the same regard for accuracy and correctness of detail to his work on Captain Cook; never being satisfied to accept what had already been written about him, but always returning to the source, or to Cook's own journal. In so far as it is possible two hundred years on, we have tried to do the same.

Our second source was *The Life of Captain James Cook* by J. C. Beaglehole, without whose life of dedicated scholarship little could have been achieved in any modern study of Cook, for it was he who also edited the magnificent four-volume edition of the *Journals*, from which, after a thorough reading, we have drawn our third and main source, – James Cook himself.

Throughout, it has been our aim to let James Cook, the man, speak for himself. We did not wish to intrude upon him with our own comments, and have therefore tried to keep them to the very minimum, for we feel most strongly that what he said is the important thing; not what we think, – or anyone else for that matter.

It is not so much that we are deliberately ignoring what scholars have written, but that we feel it is time to take another look at what the man himself said. We hope that this book will be a finger post which will send the reader back to the real source: James Cook, a skilful and graphic writer.

Tom and Cordelia Stamp

Whitby 1978.

Note from the distaff side.

There have been books upon books about Cook. There is even a book about the books. Men have analysed his reasons for doing This, his motive for doing That. They have charted his voyages and worn his pioneer steps into a beaten track.

No one, to my knowledge, has ever looked at James Cook from a woman's point of view. I find this strange, for to me he is the essence of a woman's dream of manhood; sturdy, courageous, undaunted, — yet kind and just withal; sailing away to do brave deeds, but not in battle, — it was not for him to gain so called glory by taking other men's lives.

And who would not be happy, married to such a man? Firm, yet gentle, loving children and enjoying playing with them; he was by no means stern. Mrs Cook did not care for the Webber portrait, for she said it made him look stern, which he undoubtedly was not.

Mrs Cook in her turn gave him all that a man wants: Love, welcoming warmth, security, adoration and above all, fidelity. She was utterly faithful to his wishes. She carefully destroyed every personal letter, well knowing that he did not wish their intimate life to be exposed to every eye. Mr Cook's public face was one thing, his private life at Mile End was quite another.

That they were happy together is beyond question. The six children bear witness to the fact that it was more than a marriage of true minds. Her fidelity to his memory makes me warm towards her.

Every woman needs a hero. James Cook is mine.

Cordelia Stamp.

CHAPTER 1

CLEVELAND CHILDHOOD

> You boys and girls of Cleveland
> Lift up your hearts on high!
> These noble hills about you
> Sweep upward to the sky ......
> So may your aspirations
> For ever upwards climb,
> And fill with high endeavour
> The fleeting glimpse of time.
>
> *Edward Appleyard.*

"God send you Grace!" were the words with which the mother bade farewell to her son as he left the small village of Ednam to seek his fortune. Ednam lies a short distance from Kelso, just over the English border, and thus James Cook senior could rightly claim to be a Scotsman, though his dialect led him to be taken for a Northumbrian.

Though he hardly found a fortune, he undoubtedly found Grace, for that was the name of the girl he married at Stainton-in-Cleveland in 1725. He worked as a labourer on a farm at Marton-in-Cleveland which belonged to a Mr Mewburn. They lived in a small thatched cottage, at one time the village ale house known as the *Bear* and kept by William Pearson. It was here, on 27th October, 1728, was born their second son, James Cook, destined to cross the world and venture as far as any man might.

The tiny dwelling was demolished in 1786 by one Major Rudd, who busied himself building a fine hall and 'laying our pleasure grounds' and generally undertaking some eighteenth century development with about the same regard for history and posterity as that shown by many present day 'planners'. A willow tree was planted at the spot where the cottage had stood, but not one brick remained when it was visited in 1816 by George Young, from whose biography of Cook much of this book is taken.

Major Rudd's building was burnt down, – by the angry fates,

1

one would like to say — not fifty years after he had built it.
Then in the middle of the nineteenth century a wealthy
ironmaster, Henry Bolckow, built Marton Hall there and placed
a granite vase to mark the site of the original cottage. Later,
the grounds were given to the town and remain today as Stewart
Park, the green lung of the industrial town of Middlesbrough.

After a few years on Mewburn's farm Cook senior went to
manage a farm belonging to Thomas Skottowe, Lord of the
Manor of Great Ayton, a few miles from Marton. James Cook's
father was no ordinary man. Though he has often been dismissed
as 'farm labourer' or 'hind' these descriptions, wholly inadequate
and misleading, give us no insight into the character and
capabilities of such a man. The modern term of farm manager
is more to the point and certainly more adequate than the
older title. Apart from any other consideration, the father of
James Cook could by no means be other than capable, for we
cannot believe that any gifted person came from commonplace
parents. The skills and knowledge necessary to farm successfully
are extensive and various and cover a wide range of experience.
The English weather sees to it that no man can farm for long
without becoming a philosopher.

The charming, quiet little village of Ayton, with its two
greens, both owned by the Lord of the Manor, and the fast
flowing (sometimes, alas, fast flooding) river Leven, lies beneath
the splendid line of the softly sloping Cleveland hills, dominated
by the peak of Roseberry Topping. Employment there was
mainly agricultural but, strangely enough in a village some ten
miles inland, in the last century and perhaps even in Cook's
time, there was a colony of sea-faring folk. Old people in Ayton
in the first quarter of this century could remember tales, handed
down from former generations, of the Ayton Mariners setting
off down Green Lane bound for Cargo Fleet, long years before
Middlesbrough was anything more than a hamlet.

Skottowe was a worthy, well respected country gentleman
who lived at Ayton Hall. As Lord of the Manor he presided over
the manorial courts, held from time to time to dispense local
justice. One of his farms was called Aireyholme, and it was here,
three miles away from Ayton, that James and Grace Cook lived,
farmed and brought up their growing family. There were nine
children altogether, though several died in infancy.

The farmstead of Aireyholme has to this day a character and
style much as it did in the days of James Cook's boyhood there.

AIREYHOLME. The steep road winds up to the old farm where little has changed in two hundred years. Though electricity has made it less arduous, the farming day still starts at dawn.

The Cleveland hills form, as of old, the view, with the nearby Roseberry Topping prominent against the sky. The landscape of bold and beckoning hills was well fitted to stimulate the adventurous longings of such a boy as James.

The ancient pack-horse road which runs over the moors and leads eventually to Staithes, passes by the door of the solid Yorkshire stone farmhouse. Climbing up the long steep road from the village to the farm, one would not be surprised to meet Farmer Cook with the lad James by his side. They would not seem strange in such a setting, so little has it changed since those times. In spring time there are larks singing and primroses along the banks beside the sloping track which leads up to the farm. A turn in the road brings Roseberry Topping rising into

view. Soft showers leave the air filled with scents born of field
and farm. With milking machine and tractor mercifully silent it
is not difficult to feel the spirit of the Ayton adventurer over
wide seas which dwells there.

To the imaginative mind there is always the spirit of a place
associated with great men. At Aireyholme, with the everlasting
hills around, a little quiet reflection brings Cook very close. We
see him now, striding along the three miles from school to home.
Who knows what plans are revolving in that active brain? For
the legs are the levers of the mind and many a great idea and
lofty resolve have been born during solitary walks. The walks of
boys who grow into men like James Cook are never as the aimless
meanderings of lesser spirits.

The foundation of Cooks's robust health and vigorous
constitution was laid in these Cleveland farming years. Food
was scarce, especially during the winter months; the diet was
monotonous, though wholesome and nourishing. There was no
room for fads at the big farmhouse table; those who refused
anything were bluntly told: 'All the more for the others.' The
children very soon learnt to eat all that was put in front of
them. Time and again Cook's mother would cast an anxious eye
on diminishing stocks as the winter wore on, though like many
another Yorkshire farmer's wife, she was a skilled manager and
good provider.

James Cook's ability to absorb the coarsest of food stood him
in good stead in his later years of adventure. His fellow officers
later commented on his extraordinary digestion, and it was
those early years of frugal fare that helped to develop it.

Thomas Skottowe, only a couple of years younger than Cook
senior, was a benevolent, kindly master and good to his tenants.
It was not long before he noticed that his hind's son James was a
bright little boy. He showed a kindly interest in him and arranged
for him to attend the village school. He had been taught to
read by a Mrs Walker in his Marton years. Under the guidance
of the Great Ayton schoolmaster, Mr Pullen, young Cook
studied in the little village school and showed a special
aptitude for mathematics.

There can be no doubt that Cook's career would have
developed much more slowly and perhaps taken a different
course, had not this early opportunity presented itself, owing
to the benevolent interest of his first patron and friend. The
importance of such early educational opportunities can hardly

be exaggerated. The little schoolroom at Great Ayton was for Cook the entrance to a new world that was, in the fullness of time, to extend to the uttermost parts of the earth.

Skottowe's interest in the lad did not end there, but followed him right through his life. He was a man of some standing and weight in the county, and many years later it was he who was instrumental in persuading the local member of Parliament to write to the Admiralty recommending Cook. Influence was all important, and Cook was fortunate to find such a kind patron so early in life.

Aireyholme farm lies some three miles from the village of Ayton and opportunities for play with school fellows were very infrequent. From sunrise onward the farming day brought a continuous round of jobs to be done, with no machinery or plumbing to lighten the load. All members of the family had their duties, which would admit of no backsliding. Since schooling was regarded as a luxury, James would have to complete his daily tasks before and after school. Nevertheless, in after years old men in the village who had been boys along with Cook readily recalled that the boy James had always been the undisputed leader in the occasional gatherings of the local lads, — for they also had their work to do. They remembered his unshakeable faith in the superiority of his own plan over any other, and how he always held fast to it. This might have been regarded as obstinacy, were it not for the fact that his was invariably the best scheme and was usually accepted as scuh by the others.

A story is told of how James Cook once led the way up the sloping eastern ascent of Roseberry Topping, the highest peak in the North Riding of Yorkshire. After they had quenched their thirst at the Fairy Well near the summit, he pointed out and named all the towns and villages that could be seen from there. (In later surveying years almost always the first thing he did was to climb to the top of the nearest hill.) He refused to go down the accepted way with the other boys, but boasted that he would be first down. Seeing a jackdaw fly into a cleft, he followed it up the rocky crag and took some eggs from its nest; then he had to come down, but this was not so easy. He put the eggs in his cap and held it between his teeth. The sapling which he held on to started to come away by its roots. Cook gave an agonised cry and fortunately for him it was heard by a sentry posted at the beacon at the top, for 'the meditated

ROSEBERRY TOPPING. The duck pond at Aireyholme farm stands much as it did when James Cook was a lad there. In the distance Roseberry Topping still calls to the adventurous heart and the larks sing in the quiet country air.

invasion of Charles Stuart had alarmed the country.' The boy was rescued just in time and the account (given by John Watkins in 1837) concludes: 'He retained his courage' (and presumably the bird's eggs) 'but tempered it with more prudence in future.'

School years were followed by a period of working on his father's farm. Then, at the age of sixteen, he went to work in Staithes.

The clue to the reason why it was to Staithes that he went has only very recently come to light. If his parents had merely wanted the lad to 'better himself' by going into trade, then Stokesley or Guisborough, two nearby market towns, would have been the most likely choice. There were shops enough there. But no, it was Staithes, a good two hours' journey by pony and more on foot, where Cook went to William Sanderson's shop.

Sanderson is an Ayton name. We find it both before and after Cook's time. In the last century there are many records of worthy members of the parish bearing the name of Sanderson. The link between William Sanderson of Staithes and the Sandersons of Ayton has not yet been discovered. The task today is well nigh impossible, for the parish records were so badly damaged by a disastrous flood that they became quite useless.

Documentary detectives have been busy trying to find the connection between Skottowe and Sanderson. That there is a connection is certain, for Skottowe left him two hundred pounds in his Will, a considerable sum for those days. Whatever the reason may be, it certainly adds weight to the theory that it was through Skottowe's influence that James Cook went to work for Sanderson. Ayton people, who have had the oral tradition handed down from generation to generation have known all along that it was Mr Skottowe who was Cook's patron and that but for him Cook might never have done what he did. They also know that his interest did not end when Cook left Ayton.

Undoubtedly, Thomas Skottowe, Lord of the Manor of Great Ayton, had some connection with William Sanderson, and it is therefore understandable that he should send his hind's son to work in the Staithes shop.

Staithes was a busy, thriving fishing town of some thousand inhabitants which nestled beneath the high cliffs of the North

Yorkshire coastline, ten miles north of Whitby. The greater
part of north east Yorkshire was supplied with fish from Staithes,
where there were fourteen five-men boats and about seventy
cobles, — twice as many as Runswick and Robin Hoods Bay,
their only serious rivals in the trade. Whitby did not then bother
itself much with fishing, not inshore at any rate. The boats they
built were mostly larger craft and used on the whole in the
coal trade. The whaling trade had not reached Whitby when
Cook first went there, 1753 being the year when the first
whalers sailed from Whitby.

In a small fishing town like Staithes the natural world shaped
directly the lives of the inhabitants. The world of wind and
waves, of sea and sky and changing seasons, the mid-eighteenth
century world of a small fishing and seafaring community was
more directly linked to nature than urban dwellers of the late
twentieth century can possibly imagine. This world of small
boats and great seas was Cook's world and he glimpsed it first
and with never to be forgotten power from Staithes.

Sanderson's shop was on the sea front. It has long since fallen
into that very sea which ceaselessly makes inroads into the
Yorkshire coastline, causing the inhabitants to make strong sea
walls to withstand the battering of the waves. From the shop
Cook could watch the busy North Sea, — the 'German Ocean'
— seldom without a white sailed ship passing up or down the
coast.

It was no longer work on the farm and birds' nesting that
occupied the leisure hours of James Cook. His Staithes friends
were of the sea, not of the land. The bigger fishing smacks went
off to the fishing grounds for the best part of a week, sailing out
of the little harbour on Monday and returning with the catch in
time for the Friday market. There was a large Catholic
community in the area and consequently there was always a
demand for Friday fish. The cobles, however, did not travel as
far afield as the big boats, and it was on these that Cook first
learnt the rudimentary skills of navigating.

Smuggling was rife. The State had not the firm hold on the
nation's income that it has today, and did its best to collect the
taxes levied on most imports, notably spirits and 'baccy'. It had
its 'preventive men' all along the coast, but they could not be
everywhere at once. They were by no means as numerous as the
smugglers. With over seventy little boats going to and fro,
ostensibly fishing, what could a handful of preventive men do?

STAITHES. From an early postcard. The inaccessibility of the small fishing port can easily be seen. The place where Sanderson's shop stood was in the centre of the picture, to the left of the slipway.

The fishermen were united to the last man against the common enemy, for when has a taxgatherer ever been a popular figure?

It was at Staithes that James Cook first learnt the art of quietly bringing a small boat inshore and navigating in the dark with absolutely no sound at all; an art which was to serve him well when he later sounded the St Lawrence river right under the nose of the French enemy.

Sanderson's shop stood next door to the present *Cod and Lobster* public house and on the beach below, at low tide one can still see the cobbles which formed the shop's original foundation. It has come down to us as a 'Haberdasher's' shop, a word now almost out of date. But it was much more than that. On one side was the grocery section with open sacks of sugar, flour, dried peas and beans, their edges neatly rolled round at the top, ready to be weighed out into little blue bags on the shining brass scales; sides of bacon awaited the skilled knife of Mr Sanderson, while slabs of butter, cheese and lard stood ready to meet the customer's demands. The whole shop was pervaded by that indefinable mixed spicy grocery smell based on candles and sugar. On the other side of the shop the

customer could buy all she needed for her sewing, — buttons
and thread; dark blue wool to knit the men's jerseys; canvas, to
make the proggy rug out of clippings; dress materials, pretty
cotton to make into sunbonnets. These sunbonnets are one
quite delightful legacy of the past which has survived until today
in Staithes. One may still see them proudly worn there, all year
round, regardless of sun, indicating that the wearer is a 'Steers'
woman, to use the local word for Staithes.

Few people who have not worked in a small town shop can
have any idea of just how small and how trivial that life can
become. Most customers are usually absorbed by their own
immediate needs. Perhaps the most trivial of all is the shop-
keeper's own absorbing interest in his particular small trade.

Cook's experiences among the Staithes fishing community
while he was working in Sanderson's shop had a vital, directive
influence in shaping his career. These adolescent experiences
turned his mind irrevocably towards wide horizons and distant
shores and made the limited, not to say petty, confines of a
small town shop intolerable. He soon found that life behind a
shop counter was not for him. Matters came to a head after about
eighteen months. A customer paid for her goods with a bright,
newly minted shilling. Curious, and anxious to have the coin
for himself, the boy put a shilling of his own in the till and
took the shiny one. But his master had noticed the coin earlier,
— it was a *South Sea* shilling and had *SS* on it, and when he
came to cash up he missed it. He suspected Cook of stealing it
and this it was that tipped the scale for Cook and brought into
perspective the preoccupation with trivia which life in a small
shop entails. He determined to leave and told Sanderson that he
wished to go to sea.

William Sanderson had only taken the lad into his shop on
trial and not as a regular apprentice. He had an affection for
him and a kindly interest in his affairs. He went with him to
Whitby and introduced him to his friend, John Walker.

CHAPTER 2

WHITBY

The town where Cook came down to the sea
To learn his seaman's skill.
Where he kindled the flame of his splendid fame
That burns more brightly still.

When Cook came to Whitby, he came to a busy seaport where ship building, sail making, rope and cordage manufacturing and all the allied trades flourished. There was a continual stream of water-borne traffic and hardly a moment in the day when no sail could be discerned from the pier head, there were ships crossing the bar, bustle and activity associated with docking and landing, ropes flung, capstans whirring and the creaking of sails being lowered, men shouting, and everywhere the sound and smell of the sea and sea-going life that was the dominant activity of the old town.

Writers not acquainted with the old seaport of Whitby have tended to under-estimate its influence in shaping the career of Captain Cook. Those steeped in the history and tradition of the place know better. This ancient town, with its monastic and maritime associations stretching back to earliest times was the very port exactly fitted to capture the imagination and foster the ambitions of such a one as Cook.

The forty-year-old John Walker, to whom the young Cook was apprenticed, owned, with his brother Henry, several ships. The Walkers were a respected Quaker family and lived in Grape Lane. The house still stands today and the visitor will see the initials of Mark and Susanah Dring, who first built it in 1685, on the wall outside. The back of the house looks out on the river Esk and at the top the large attic room remains much as it was at the time when Cook, according to long standing Whitby tradition, slept there. There, one can hear the ceaseless sound of the wind singing in the eaves, mingled today with the noise of traffic. Some confusion has lately arisen over the locality of the house where Cook's master lived. This is understandable, since John Walker's cousin, some three years his junior, as well as

being named John, was also a master mariner. He lived in
Haggersgate, Whitby. We have not found any firm evidence that
Cook ever stayed there.

The eighteenth century Quakers were a body of people much
more aware than most of moral and spiritual issues and they
strove to work out in daily life their inner convictions with an
earnestness beyond any other section of the community. *Advices*
were issued from their London Yearly Meeting on every
imaginable topic, for religion pervaded their whole, day to day
life. Though they had no rigid creed they were *advised* on all
subjects from *Conduct and conversation* to *Tithes* and *Temperance*, from *Moderation* and *Meetings* to *Mourning habits*.

Friends were 'Advised in God's holy fear to watch against
and keep out, the spirit and corrupt friendship of the world
. . . Avoid unnecessary frequenting of taverns . . . unprofitable
and idle discourses, mis-spending their precious time and
substance to the dishonour of truth.' They were 'advised to be
careful of their conduct at all times, . . . to be prudent in all
manner of behaviour; to avoid all pride and affectation and
have frequent waiting in stillness on the Lord for renewal of
strength.'

They were aware that 'much hurt may accrue to the religious
mind by long and frequent conversation on temporal matters,
especially by interesting ourselves too much in them.' A
plainness of living was earnestly recommended and usually
followed, though not perhaps as strictly in the middle of the
eighteenth century as in the seventeenth.

Quaker ship owners in Whitby were in the black books of the
weighty worthies of their London Yearly Meeting, who, in a
genuine spirit of pacifism, advised their sea-going Friends not to
carry guns on their ships. This was all very well for inland
Quakers, but Whitby ship owners had to recruit crews for their
ships and no sailor in his right mind was going to ship aboard a
defenceless vessel while there were privateers on the high seas.
And so the Whitby Quakers, the Walkers among them, continued
to carry the guns, but it tended to split the meeting. Some of the
members were disowned and others, John Walker among them,
later went away to other parts where there was more tolerance
and less dissension.

Henry Walker, John's brother, had also been in disgrace with
weighty Friends. His name appears again and again in old Minutes,
'Hy Walker admonished', 'Hy Walker reprimanded.' His crime

had been to marry Ann Hudson *in a church*. For this 'disorderly action' he could have been disowned and put out of membership. But Whitby Friends Meeting was a law unto itself. They ignored the advice of other Quaker meetings in the area and Henry stayed where he was. All this, however, occurred some twenty years before Cook came to Whitby. By then the Walker household had increased with John and Henry's children – and cousin John's too – growing up. There were eight girls, cousins, in the younger generation, all more or less of an age with the young James Cook. The family house in Grape Lane must have been a merry place when they all gathered there.

A young man like Cook could not fail to be impressed by the Quakers' example of uprightness, integrity and fair dealing, qualities exemplified in his master, John Walker, to whom he was bound apprentice for three years.

The Walker ships were in the coal trade. They were known as 'cats', and were extremely sturdy, stubby little ships with bottoms as flat as a jellyfish, which could be easily moved close in to the wharves which lined the broad sides of the Esk at its mouth. They needed these flat bottoms in order to negotiate the shallow tidal harbours and sandy coastal waters that fringe the north easterly shore of England. It was in piloting these ships that Cook learnt his unrivalled seamanship and navigating skill.

The *Freelove*, on which he first served, was about 450 tons. There was a comradeship aboard a collier not to be found in the Navy. The men who sailed on her, unaccustomed to the discipline of the lash, were workers proud of their ship. Each man pulled his weight or was made to do so by his fellows. In port, – loading coal at Shields, or off-loading in London – it was dirty work, but once the cargo was stowed it was a good life.

James Cook soon made his mark and came under the approving eye of his master who, having no grown sons of his own (his two boys were but infants at this time), took a fatherly interest in the lad. He employed him in helping to rig and fit out a new ship, the *Three Brothers*, perhaps named after the three Walker brothers, John, Henry and Thomas. She was a bigger ship than the *Freelove*, being about 600 tons. It was Cook's first experience of such an undertaking and stood him in good stead some twenty years later. A ship that was to be away from home for a long period had to carry absolutely everything that

was needful, yet nothing unnecessary. This applied just as much
to the *Endeavour* as it did to the *Three Brothers* in 1747.

When the winters were severe, as they can be on the North
East seaboard of England, the colliers were laid up at Whitby
and were thoroughly overhauled and cleaned in readiness for
the next year's work. Thus it was that, — twenty years on —
Cook was familiar with the external structure of a vessel. Holed
and grounded on the Barrier Reef, he was able to draw on his
early experiences when he had become closely familiar with a
ships's construction and maintenance.

At these times it was the accepted practice for the 'prentice
lads to stay in their master's house, and it was in the Walker
house in Grape Lane that Cook spent the winter months ashore;
not in idleness, for there was work enough on the ship by day
and in the evening he studied navigation, nautical law and, say
the books, astronomy. But even before that he looked upwards
through long night watches aboard ship. He watched the stars
and little by little he learnt their names and impressed the map
of the heavens upon his retentive mind.

On the ships' muster rolls of the time Cook is called a servant.
Also called a servant was Mary Proud, — Molly, the benevolent
friend, nurse, housekeeper and mistress in all but name of the
establishment in Grape Lane. Let who dare challenge her
authority. From the very start James was one of her favourites.
She allowed him an extra candle and a table all to himself where
he could work of an evening before he went up to his hammock
under the attic beams, where slept all the lads. They were allowed
only one candle among them while they frittered away their
free time; they played cards and spun yarns and discussed the
lasses, unless they were any different from their present day
descendants, which we doubt.

Years and years later, when Captain James Cook had become
an international figure, the 'celebrated circum-navigator' was
expected in Whitby. Molly was carefully schooled. She must
call him Captain Cook and say 'Sir' when he addressed her. She
promised to do this, awed by so great a personage. But the
minute she saw him she forgot all this. She opened her motherly
arms wide, folded him within them, and exclaimed: "Oh, James
honey, Aa's glad to see thee!" He had not changed one bit.

Cook had only the beginnings of education at the village
school, but it was a good beginning and a firm foundation.
Nevertheless, we can only marvel at the splendid structure he

erected on it. Without much help from anyone except a few interested friends, he studied algebra, spherical trigonometry, navigation and astronomy. The capacity of his mind and his ability for intense application to study was remarkable and especially outstanding for one with his background and, later, busy seafaring life. Those winter evenings passed in the glow of Molly's candle flame were wonderfully well spent, as were the later winters in Newfoundland. At these times Cook mastered the intellectual tools of his trade and later, around the oceans of the world, used them to superb effect and in a way none had done before him.

Much has been written about Cook the explorer and navigator, but the scientific basis for his achievements acquired by an outstanding effort of self-help in education is of considerable importance in understanding his character.

During Cook's service in the *Three Brothers* she was commandeered and used as a troopship to bring men from Middleburg, in Holland, after they had served in the war of the Austrian succession. They were taken to Dublin, where the position was little better than it had been for centuries. Once again, Cook had a valuable experience which was to serve him later in life. He saw how the human cargo was stowed away on the ship; how the soldiers messed; how they were disciplined, and a dozen other things. There were lessons there to be learnt and laid into the recesses of his mind which were of use in later years. After the ship was paid off at Deptford she was engaged in the Norway trade.

After this it is recorded that Cook served in the *Mary* which was owned by John Wilkinson. She was in the Baltic trade and her master was Captain Gaskin, a relative of the Walkers.

In 1751, Young tells us, Cook shipped aboard 'a Stockton ship' whose name we do not know. It may be noted in passing that 'Stockton' meant the River Tees. A hundred years had yet to pass before the small beginnings of Middlesbrough appeared on its banks. Stockton was the port to which ships sailed. Less than a score of years ago one could wander down narrow, tarry-smelling cobbled streets and come upon old, long forgotten grass covered wharves and jetties, not two hundred yards from the main street with its elegant eighteenth century town hall and covered market. Stockton on Tees was a busy port and a town of some standing and much wealth, as its Wren church testifies. Though the old town has almost completely disappeared beneath

the tombstones of development, one can still get something of
the feel of the old life along the riverside, from whence one can
see the distinctive peak of Roseberry Topping in the distance.
The grass grown banks are still there, and sea birds wheel about
one or two small rowing boats tied to rotting posts, while a
barge chugs slowly past. At the blink of an eye one can almost
see its ancestor, the stubby collier off-loading coal, and all but
hear the one time clamour associated with ships and ports and
dray horses coming to fetch the cargo.

Today, Stockton, — like the village of Marton — has been
swallowed up by the ever encroaching maw of the urban
conglomeration now called Cleveland County, and were it not
so sad, it might be thought ludicrous that such a tangled
industrial spread should claim kinship with James Cook. We in
Whitby feel proud that our old town has changed very little
from the way it was when Cook walked its ways.

After the Stockton ship Cook returned to Captain Walker and
for three years sailed with him as mate in the ship *Friendship*.
She was in the coal trade, so it was up to the Tyne and down the
perilous East coast to the Thames over and over again. No
journey was ever the same. If a man could sail that tricky coast
with its hidden shoals and unpredictable currents, he could
sail anywhere. Storms could blow up with very little warning and
in no time at all a ship could be driven inshore any where from
the Black Middens to the Maplin Sands.

The men who built these sturdy, flat bottomed ships knew
the dangers of the northern sea board full well, and the way
they built them gave an extra chance of survival. If they were
driven aground, they could be got off without much damage
from places where other ships would have been total wrecks.

The men who sailed with Cook were almost all Whitby men;
if they were not actually born in the town they had moved
there from outlying districts, as he had done. We see the familiar
Whitby names, — Storm, Pennock, Kirby, Dixon and many
others — alongside Cook's on the surviving muster rolls, now
in Whitby Museum but preserved for many years by the Seamens
Hospital. These mariners' almshouses are an ancient charity, a
cluster of a dozen or more dwellings built alongside the river;
not today as they were in Cook's time, for Gilbert Scott came
and put his medieval hand on them in the middle of the last
century. The foundation, however, is the same, for the dwellings
are let only to deserving seamen and their families. The original

charity levied a toll on all the ships in the port, so much for each ship, so much for the master and mate, and so much — a few pence — for each man who drew over one pound in wages. The Seamens Hospital held accurate muster rolls with details of all ships and their crews and it is from these books that we learn the names of Cook's associates.

And what of these Whitby men? They were born to a tradition of 'going to sea'. The sea dominated the lives of all who lived in Whitby for it was their means of livelihood and of communication. Landwards, Whitby is surrounded by hills. Only the sea gives level access, and it was by sea that almost all goods entered the town. Whitby people tended to have more knowledge of the outside world through the sea and their menfolk's travels over it. Inland folk were of the simplest, living and dying in the same villages as had their parents; tilling the soil from season to season. But despite his wider experience the Whitby man — and Staithes too — had his share of superstition in common with most sailors. He did not like to turn back once he had set out from home to his ship. He disliked meeting a woman on this occasion, a priest was worse, but a pig was the most unlucky thing to encounter. Indeed, he would often not go to sea if he met one. The poor old piggy is so loaded with bad luck in Staithes that he is not even referred to by name, merely as a 'porker' or 'old grunter' or some other pseudonym. Cook was not taking any chances. He almost always refers to them as 'hogs.'

There are many misconceptions about seafaring men among landlubbers. It was a hard life, let there be no doubt about that, but the picture of the hard-drinking, coarse, swearing, dissolute character so often portrayed is largely fantasy. There were some disreputable types, — all callings have some such, but the demands of the sea were stern and unrelenting and required comradeship, quick action and quick thinking, a mastery of the work to be done under difficult and often well nigh impossible conditions. Almost wrecked in a fearful storm, with unseen jagged coral reefs threatening them, Cook wrote: 'In this truly terrible situation not one man ceased to do his utmost and that with as much calmness as if no danger had been near.'

The deck of a ship in the eighteenth century was no place for a laggard and a misfit. It was the same with the Navy. The brutal, sadistic commanders we so often hear about no doubt

WHITBY at the turn of the century. The small, five man fishing boat was very little different from those with which Cook was familiar.
(Photograph by F. M. Sutcliffe, by permission of the Whitby Literary and Philosophical Society by arrangement with the Sutcliffe Gallery, Whitby.)

did exist, but they were a hateful and a hated minority. The service could never have functioned as an efficient fighting force had it been otherwise. The men of the merchant fleet, as of the Royal Navy, were mostly stalwart, tough, patriotic men of the sea. They were the type of men with whom Cook cast his lot and of whom he became an outstanding example.

Whitby is little different from most towns in claiming the loyalty of its citizens, much of which today is channelled into support of the local football team, but the people there are also bound to their town with rather more uncommon ties. In Whitby it is the sea, always the sea, which has first claim on its menfolk. It is the same today as it was in Cook's time. Talk to any native of the town and it is almost certain that there will be a seafaring link in that family. If the man is on shore now, he has probably been to sea at some time in his life, or his father or his brother went to sea. Whitby people have an affection for the place and a loyalty towards it greater than most. If the master of a passing ship happens to be a Whitby man, that ship will divert its course, swerve inland towards the town and give a friendly *toot-toot* to remind the town of its maritime connection.

Even when Cook was not actually living in the town he was with Whitby men constantly in their ships. Their ways and turns of speech were absorbed unconsciously and remained with him all his days.

It is impossible to understand Cook the man without a just appreciation of the formative influence of the years he spent as a youth and in his early manhood in and around Whitby and sailing in the North Sea coastal trades. Many boys born into poor homes on the northern seaboard have in time become master mariners. They have obviously possessed intelligence, determination and the will to succeed. Cook was of this type. Where he differed from his fellows was in the magnitude of his gifts. His qualities were their qualities, but much enhanced. Thus, today, if we meet one of these sturdy, tactiturn yet bright eyed sea-going men of the world, we stand in the presence of a man not so very different from James Cook.

CHAPTER 3

THE ROYAL NAVY

I vow to thee my country,
All earthly things above,
Entire and whole and perfect,
The service of my love.

Cecil Spring Rice

Cook enlisted in 1755 for precisely the same reason as did thousands of men and women in 1939. It was not the thought of rapid promotion. It was not to avoid being impressed, or to be better thought of as a volunteer. It was for a multitude of reasons, but the predominant one was that he was British and his country was in difficulty. They needed men and they needed men skilled in their craft. Cook was one of these; his years in the Whitby ships had given him experience second to none. 'Your country needs you,' they said, − both then and now.

It was in response to this call that James Cook, mate and master-elect of the *Friendship*, went to the *rendezvous* at Wapping on a spring day in May 1755. Though war was not declared until the following year, relations had been extremely uneasy between the French and British in North America and hostilities had actually broken out. Early in 1755 a British fleet was sent to the American station with orders to attack any French squadron which might be found in those waters. Hearing of this, the King of France, Louis XV, said that the first shot fired by the British would be taken as a declaration of war. The shot was fired on the 6th of June, but the official declaration did not come until the following May, 1756.

In any given situation the great majority of people tend to make exactly the same comment and invariably ask the same questions. In after years, when people asked Cook why he had joined the Navy, thinking he must have been mad, conditions in his day being as they were; thinking *they* would not have done such a thing; thinking *they* would have accepted the proffered command of the collier and travelled up and down in well defined tracks and charted waters, − Cook knew their thoughts and produced a stock answer that would satisfy them all. "I had a mind to try my fortune that way," said he, firmly silencing those landlocked critics. He was indeed 'a plain man, zealously exerting himself in the service of his country.'

The new recruit was sent to join the *Eagle* under Captain Hamer. In a very short time his worth had been recognised for he was promoted to master's mate, under Thomas Bissett, the master. The ship went a short cruise in St George's Channel between the Scilly Isles and Cape Clear in southern Ireland and chased various ships thought to be French without success, though it was a splendid opportunity to press men from the smaller craft which they encountered. On the *Eagle's* return to Plymouth, Hamer was succeeded by Captain Hugh Palliser. Cook's log reports: '1 October 1755. Came on board Capt. Palliser & took possession of ye ship.'

Captain Palliser (later Sir Hugh Palliser), a fellow Yorkshire-man, was only five years Cook's senior yet he had already had twenty years naval service, having gone to sea as a boy under the care of an uncle. He was a first class officer and Cook was indeed fortunate to serve under such a man. He was one of those rare spirits capable of discerning ability in others, and large minded enough to acknowledge it and to promote it wherever possible. He remained ever afterwards a patron and friend of Cook.

For about two years the pattern of Cook's life was much the same. The *Eagle* cruised in and out of Plymouth Sound looking for French ships, for England was now officially at war with France and they were trying to stop supplies going to the French in Canada. Down to the Bay of Biscay and up to the German Ocean Palliser's ship scoured the sea for prizes and had several encounters with the enemy.

It was on these runs that Cook first met the sailor's greatest enemy, scurvy. Today we calmly dismiss it as a vitamin deficiency disease. We know that what they needed was vitamin C. But they did not know this; it was to them as much an unsolved mystery as cancer is today.

It starts in the gums, with swelling and soreness; blisters soon develop until eventually the teeth start to drop out. The disease quickly runs through the whole system, causing bleeding and swelling both inside and out. Extreme lassitude sets in and movement of any sort becomes very difficult. Then comes delirium and convulsions and finally fever takes over. Scurvy was a dreaded scourge and the Navy lost more men that way than they ever did in battle.

Though we are not really certain of the result, we do know that while Cook was serving on the *Eagle*, Captain Palliser received a letter from William Osbaldeston, the Member of Parliament for Scarborough, recommending Cook. The letter was written at the instigation of Thomas Skottowe, Lord of the Manor of Ayton, and John Walker, the Whitby Quaker ship owner. From this distance of time we like to think that Cook did not need any recommendation and that his own worth was enough. But this was not the case; it is very necessary to transport ourselves back to the middle of the eighteenth century, to the first years of the reign of King George the Third. In those days influence played a very large part in shaping a man's career and the letter from Osbaldeston could only do good for the young Cook.

What is most remarkable is the fact that persons of importance and influence were exerting themselves on behalf of a member of the lower deck, – a seaman, a person of no connection and no standing. We know full well that all sorts of people of wealth and standing sought to exert their influence on behalf of 'young gentlemen', relations and younger sons of the gentry, but to have pressure brought to bear on a commander on behalf of an obscure seaman was very unusual.

Shortly after this letter Cook was promoted to the rank of Master. The position was a most responsible one and carried considerable prestige, though the holder was of inferior social standing to the officers. The master was appointed by warrant from the Navy Board and not by commission from the Admiralty. He was the chief professional on board responsible for navigation. He was over the boatswain for stores, masts, yards, sails, rigging and general management. The master had special responsibility for pilotage and harbour work; for taking soundings and bearings. He kept the ship's log. In fact, his duties were endless and his oversight was everywhere.

The position had its roots in the historical situation when kings and princes and aristocratic personages, armoured and armed, clanked aboard some requisitioned ship and set sail to fight at sea, leaving the tasks of navigation and ship management to the master. Such menial tasks were, of course, well beneath their notice and certainly well beyond their competence. The present day use of the title *Master Mariner* has its honoured and honourable roots in this historical fact.

Cook's first appointment as Master was on the *Solebay*, which he took up on patrol off the east coast of Scotland as far as the Shetland Isles, looking out for smugglers and those having 'treasonable intercourse' with France. That smuggling did go on, despite the war, can most certainly be inferred from a Quaker Advice of that time: 'It appears that some members buy run goods for their own use . . .' being staunch pacifists, however, they were not so concerned with patriotism as with the bad effect on trade: 'Dear Friends, surely you don't sufficiently consider the pernicious consequences such conduct is of to the fair trader?'

The anti-smuggling run did not last long for Cook, for within three months, on his twenty-ninth birthday to be exact, he became Master of the *Pembroke*. This was a 24 gun, 1250 ton naval ship and not to be confused with the *Earl of Pembroke*, less than 400 tons, the Whitby ship whose name was later changed to the *Endeavour*.

Cook's new captain was John Simcoe, 'a truly scientific gentleman', and credit is due to him for encouraging Cook in his nautical studies, for it was he who later supervised these studies and gave him text books.

For her first winter, − for the *Pembroke* was a new ship − she cruised south and around the Bay of Biscay looking for French ships in order to stop them taking supplies to Canada. At times she chased them alone, at others in company with the whole fleet. Cook took part in naval action and watched enemy ships going down. But the real theatre of war was on the other side of the Atlantic and Britain was preparing for action there.

At the end of February, 1758, the *Pembroke* left Plymouth in company with eight line of battle ships, the *Eagle* and Captain Palliser among them, and a number of transports, bound for the scene of trouble. They arrived in Halifax in early May, but the *Pembroke's* crew was sadly depleted. Twenty-six men had died on the passage, a great number had to go straight into hospital and, as if that were not enough, five men deserted, taking the ship's yawl with them. So the *Pembroke* could not join the fleet at once, but was delayed a month. When she did join the rest of the ships, Cook then saw full scale naval action and fierce, unremitting bombardment such as he had not before witnessed. Chasing individual ships was one thing, bombarding a fortified town was quite another matter. The fleet kept up its assault on

Louisburg until, after five weeks of siege, the French finally surrendered it.

The day after the surrender, Cook went ashore at Kennington Cove. It was here that he met Samuel Holland, a young Army surveyor (he was the same age as Cook) who was taking measurements with a plane table. Cook told his captain of the encounter and Simcoe invited Holland aboard. This was the start of what could be described as Cook's university studies.

The great cabin of the *Pembroke*, 'dedicated to scientific purposes and mostly taken up with a drawing table' was where Holland and Cook, under the watchful eye of Captain Simcoe, pursued their studies. Here Cook studied trigonometry, navigation, algebra and Euclid. Among the books which his kindly captain gave him was Leadbetter's *Young Mathematicians's Companion,* a 'Compleat Tutor to the Mathematicks'. The young Cook was fascinated with Leadbetter's book. It opened a door on to a new world which he had glimpsed but within which he could find no mentor or guide. Leadbetter became his friend and companion and gained a pupil of whom he could well be proud. It can certainly be asserted that the book had no more avid reader, nor one who mastered it more thoroughly and later applied it more successfully than James Cook. To the lone student struggling to master a difficult subject, the right sort of book is most important. Indeed, it is indispensable.

Cook studied throughout the ensuing winter while the ship was laid up at Halifax, waiting for the spring before moving in to attack Quebec. Not until May was the ice sufficiently melted to allow the ships which had wintered in Halifax to move out.

Captain Simcoe had been unwell, having to keep to his cabin for some time, and on the 15th of May 1759, he died. His friend Cook had the sad duty of conducting the burial at sea, firing 'a salute of 20 guns, half a Minute between each gun.'

By early June 1759, the fleet had reached the beginning of the Traverse, a part where the great St Lawrence river starts to narrow. It was quite impossible to navigate this part of the river without charts and the British had none. Only the French had them. For an accurate chart of the unknown river to be negotiated it was necessary not only to take surveys but also soundings. That is to say to drop a weighted line in to the water and measure, by means of the regular knots in it, the number of fathoms to the bottom. On the weight itself was some soft wax, which picked up some of the material which lay on the bottom and gave an indication of its composition. Throughout Cook's journals we find such entries as: 'Six fathoms, muddy bottom' or: 'Twenty Fathoms, sandy bottom with shells.'

The St Lawrence Traverse had to be sounded so that depths would be known. They had to know where lay the sandbanks and where the rocks. But this could not be done in a small boat in broad daylight; they would have been a sitting target for French snipers. The job had to be done under cover of darkness. We are told that Palliser still in command of the *Eagle* with the Newfoundland fleet, recommended Cook for the job. And who better than the man who had spent part of his youth at Staithes, where smuggling was rife? None knew better than he how to guide a boat in the dark, silently and quickly, with barely a splash from its muffled oars.

The Master of the Pembroke could easily have given the task to one of his subordinates, but he had to know, he had to be sure, and he knew that if he did the job himself the resultant chart would be as accurate as he could make it.

Young tells the story in words which have not been bettered:

> It was necessary to take the soundings in the channel of the St Lawrence directly in front of the enemy's entrenched camp at Montmorency; and Cook being recommended for this service by his friend and patron, Captain Palliser, performed it in a manner that gave complete satisfaction to his officers but with no small peril to himself. For several nights in succession he was employed in taking the soundings and making a survey of the channel, but when he was finishing his task, he was discovered by the French, who collected a number of Indians in a wood near the river side, where they launched their canoes to surround him and cut him off.
>
> He had just time to escape, by pushing ashore on the island of Orleans, near the guard of the English hospital; abandoning his boat to the Indians, who entered it at the stern while he was leaping out at the bow. By that time, however, he had so effectually accomplished his work that the draft of the channel and soundings was found as correct and complete as it could have been made in daylight.
>
> In executing such a task in the night time and in the

immediate presence of an enemy, Mr Cook gave a
most striking proof of his cool bravery, accurate
observation, unwearied patience and indefatigable
perseverance; qualities for which he was ever after
distinguished.

By September, 1759, the opposing forces were ready and the
attack on Quebec was made. The French held grimly on and
the bombardment was tremendous. Twice the city was set on
fire. General Wolfe gave his orders. Cook records them in his
log: 'At midnight all the row boats in the fleet made a feint to
land at Beauport in order to draw the enemy's attention that
way to favour the landing of the troops above the town on the
north shore, which was done with little opposition. Our batteries
kept a continual fire against the town all night. . . . At 10 the
English Army commanded by Genl Wolfe attacked the French
in the field of Abraham behind Quebec and totally defeated
them.'
A few days later the city capitulated and the British moved
in and took possession of Quebec.
The news took some time to reach England, but when it did
the whole country rejoiced. Great Ayton was no exception and
there was a ringing day and the bells pealed from morning to
night. The Churchwarden's book records the cost:

For ringing on account of Good News 1/–

and this is followed by a significant entry:

For Mending Bellrope 6d.

CHAPTER 4

NEWFOUNDLAND

'We can hardly tell when we are possessed
of a good sea chart until we ourselves
have proved it.'

Only a few days after the fall of Quebec Cook was given further advancement. He was appointed Master of the *Northumberland,* a larger ship than the *Pembroke* and the flagship of Lord Colville, who became another of Cook's kindly patrons. It is remarkable that every person under whom Cook served became his friend. Petty jealousies must have existed, yet Cook's nature was such that he won respect wherever he went.

The winter of 1759 was spent once again in Halifax. Although quite shut in by ice and therefore made inaccessible, the town itself was pleasant and comparatively sheltered. Cook returned to his studies and 'thoroughly mastered Euclid.' He was the same James Cook who had studied in the Walker house in Whitby while the others idled their time away. He did not waste his time in Halifax either, for he occupied it in making charts of the harbour there. Three of these remain today, and one can but marvel at the care and exactitude with which they are drawn.

The charts of the St Lawrence which Cook had already made were published, and for more than a hundred years after they remained as the standard guide of that dangerous river. There was much to be done over there; there was still no accurate chart of the coastline of Newfoundland.

That the Admiralty was becoming aware of Cook's value is apparent in Lord Colville's payment to him, in January 1761, of a bonus of fifty pounds 'in consideration of his indefatigable Industry in making himself Master of the pilotage of the River St Lawrence &c.'

James Cook stayed with the North American squadron until October 1762, when the *Northumberland* sailed for home. He landed in November and was married in December. At first, one is a little surprised at this; but when we remember that Cook was a man of action, that he was but the man grown from the Ayton boy who knew exactly where he was going, the boy who stuck to his own plan once he had determined it, then perhaps it is not so surprising.

In a few short weeks James Cook won the twenty-one year old Elizabeth Batts of Shadwell, London. Who could resist such a lover? The glory and the greatness lay in the future and were all unknown, but who can say what aura of distinction hung about his tall figure and won the love of Elizabeth in a breath-taking and altogether overwhelming fashion? Nor can she have been just any woman; to have gained the love and esteem of such a man as James Cook was no small distinction.

He, man of decision as he was, saw no lawful impediment to warrant delay and so they obtained a special licence and on the shortest day of the year, 21st December 1762, James and Elizabeth walked together over the fields to St Margaret's, Barking, and there they were married.

They were able to remain together until the following April, when Cook was given a new appointment. This was at the instigation of Captain Graves, who knew Cook and his fine work so well. A letter of his to the Admiralty makes us realize that though times may change, men do not:

> I have this moment seen Mr Cook and acquainted
> him he was to get himself ready to depart the moment
> the Board was pleased to order him. He has been to
> enquire for a draughtsman at the Tower, but as this is
> a Holiday he found hardly anyone there.

Cook was appointed Marine Surveyor and returned to the North American coast with Graves, who had been made Governor of Newfoundland. The British were anxious to have accurate charts of the islands of St Pierre and Miquelon before they were finally handed over to the French under the settlement terms. But the ship from England did not even arrive in Newfoundland until after the official handing-over date. Somehow the French had to be delayed until the vital survey was done. Between them, Captain Graves and Captain Douglas of the *Tweed* contrived to occupy the attention of the newly arrived French Governor of these islands, and he was entertained lavishly on their ships. All the while, Cook and his two assistants, Biddon and Flower, worked away non-stop at surveying the islands. It took them all of five weeks and by the end of that

time the French Governor's patience was wearing very thin.
But when the islands were finally taken over by the French,
the British had accurate charts of them, entirely due to 'the
unwearied assiduity of Mr Cook', as Douglas put it.

Having succeeded in this manoeuvre, Graves was able to
give his attention to the purchase of a small schooner, the
Sally, 68 tons, for Cook's own use. The name was changed to
Grenville and Cook took her up the coast and made further
surveys until the start of winter.

This time there was no wintering in the shelter of Halifax.
He was a newly married man and had left his wife not really
certain if there was a baby on the way. Graves understood
Cook's eagerness to be home and beneath the official wording
of his report, we detect his thoughtfulness in sending him back
to England at the first opportunity.

> As Mr Cook, whose pains and attention are beyond
> my description, can go no farther in surveying this
> year, I send him home in the *Tweed* in preference to
> keeping him on board, that he may have the more
> time to finish the different surveys already taken.

The little ship was left to winter in St John's and Cook
returned to England. Here he found that he was indeed a father,
for Elizabeth Cook introduced him to their month old son,
James. Very shortly after his return they moved from their
Shadwell lodgings to their own home in Mile End Road. It was
not very far from Shadwell and it is highly likely that Mrs Cook
had seen the house and was only waiting for 'Mr Cook' to return
before moving there. She always referred to him thus.

For the first few years of their marriage Mr Cook was at
home during the winter months, but away most of the rest of
the year. Mrs Cook, like so many wives of sea-faring men, had
to be in firm control of home and family; self-reliant, ready to
make decisions and to act on them, to plan for the home-comings
which were such an event in the family calendar. She it was who
made the little house in Mile End Road into the Cook home and
the centre and circumference of the home life of this matchless
explorer and navigator of distant seas, surely often recalled to
his mind during the long sea passages and never to be forgotten
experiences of those epic voyages.

But the long separations from wife and home and family

lay in the future and he could look forward to the winters and
the reunions, the welcome, the excitement of those times. We
can almost hear Mrs Cook saying to the children: "Wait till
Mr Cook comes."

It is indeed sad to think that, despite the blue plaque fixed
at the beginning of this century to the Cook house in the Mile
End Road, it was demolished only about twenty years ago, —
though some of the houses in the same row were spared. It is
strange that a country which has such a splendid history should
have such a total disregard for buildings connected with that
history. It is good to know that the little Ayton cottage which
James Cook's father built on his retirement will stand for ever,
thanks to the care of an Australian, Russel Grimwade, who
bought it for his country in 1934. It was transported, each brick,
stone and slate carefully numbered, and re-erected in Fitzroy
Gardens, Melbourne, where it stands today.

For five years the pattern of Cook's life remained fairly
constant. The better months of the year, — one can hardly call
them summer in that bleak, exposed, icy corner of North
America — were spent in surveying, not merely the coastline,
but the land also in Newfoundland and Labrador. Young tells
us he even discovered some coal there. There was squabbling
about the fishing rights, and Cook went among the men asking
questions and he established that the British had fished those
waters for longer than could be remembered.

Then in winter time it was back to the comparative warmth
and comfort of England and his home in London. His little
schooner, the *Grenville* was never again left to winter in
Newfoundland but, through the influence of Palliser, Cook was
allowed to bring her to England and, despite her small size,
he crossed the Atlantic in her every year. On her last crossing,
she ran into very heavy weather and Cook was obliged to throw
overboard his ballast and a good part of his heavy stores in order
to save the ship. Among the items which were lost in this way
was an Indian canoe which he was bringing home to add to the
collection of a wealthy young traveller who had visited
Newfoundland. His name was Joseph Banks.

The naval world of Newfoundland was a very small one. It
is quite possible Banks and Cook had become acquainted.
Banks, all his life, strove to discover ability in science, especially
in observing, exploring, adventurous spirits. Cook was just the
type of man to attract the young and wealthy gentleman who

GOING, GOING ... Skilled men start to dismantle the historic cottage at Great Ayton in readiness for its journey to the other side of the world.

COOK'S COTTAGE. Each stone is carefully numbered before being shipped to Melbourne in 1934.

happened to have brains and ambitions for science and his country. There is little doubt that even then he was exerting those influences behind the scenes (he was a personal friend of Lord Sandwich) that later made him such a powerful fertile figure in British science.

How was it that Cook came to be entrusted with Banks's canoe? Cook's was a very small vessel; many much larger ships would be crossing on the homeward voyage. Do we already detect an understanding between Joseph Banks and James Cook that was later to play its part in motivating the momentous voyage?

At first the *Grenville* was moored at Woolwich, but Cook managed to convince the Admiralty that she would be better at Deptford. This was very much nearer his own home. A short journey from the Mile End Road would bring him to a point on the riverside opposite Deptford, where it would be easy to ferry across. Woolwich entailed a much longer passage downstream. Cook, ever the man to conserve time and energy and to concentrate on essentials, saw all the advantages. He would not, therefore, have to sleep aboard the ship, but could return to his home with its growing family.

Little James's new brother, Nathaniel, arrived in the Cook home at almost the same time as their father returned home, on 14th December 1764. Their sister Elizabeth, the only girl of the family, was born in 1767. Mrs Cook had her husband's cousin, Frances Wardale from Yorkshire, living with her. Fortunately for the mother in those days of large families there were often spare females, – sisters, aunts and cousins, who found a useful life's work in caring for their relatives' children.

Cook's letters to the Admiralty are models of diplomacy and tact. He did not always get his own way, but he knew how to set about getting it. It may be that the years at Ayton had taught him how to be sufficiently respectful to his superiors in order to win them over to his way of thinking. He was a past master at this, though never unduly or obviously deferential. He always manages to convey the impression that the proposed scheme originated with their Lordships, and the credit for its successful completion will go to them. One example of this diplomatic approach occurs in a letter in which he asks them to alter the *Grenville* schooner into a brig.

Carefully, he sets down all the advantages of the different

rigging, 'besides many more I could name, was I not applying to
Gentlemen better acquainted with these things than myself. I
only mean to give some reasons for my request and pray you will
be pleased to take these into your consideration.' The letter did
the trick and the ship was re-rigged as he wished.

The work of charting the whole of Newfoundland went on
from year to year. An injury to his right hand in 1764 meant the
loss of a few weeks' work. He bore the scar until his death and
it was a gruesome means of identification after his murder.

By 1765 Cook had completed the northern part of his survey
as far as Belle Isle. Each year he returned to the exact spot
where he had left off the previous year and in this way the
whole area was completely surveyed. The resultant charts were
published and remained as the standard guide for well over a
hundred years, and even then they were found to be completely
accurate and reliable; which was more than could be said of
those in use before Cook, — some of which were a positive
menace to the mariner.

It was not the navigator, said Cook that was 'wholly to blame
for the faultiness of the charts, but the compilers and publishers
who publish to the world the rude sketches of the navigator as
accurate surveys without telling what authority they have for
so doing, for were they to do this we should then be as good
or better judges than they, and know where to depend upon
the charts and where not. Neither can I clear seamen of this
fault among the few I have known who are capable of drawing
a chart or sketch of a sea coast. ... I have known them lay
down the line of a coast they have never seen and put down
soundings where they have never sounded, and after all are so
fond of their performances as to pass the whole off as sterling
under the title of a *Survey, Plan &c.* If he is so modest as to say
such and such parts are defective, the publishers and vendors
will have it left out because they say it hurts the sale of the
work, so that between the one and the other, we can hardly
tell when we are possessed of a good Sea Chart until we
ourselves have proved it.'

It was while Cook was on the *Grenville* that he ran into an
old school friend. Or to be absolutely correct, the friend,
Thomas Bloyd, ran into him. It happened in the Thames, and
Cook's rage at the carelessness which occasioned the accident
in which the collier had 'carried away our Bowsprit cap and Jib
Boom' soon melted when its master made himself known as a

fellow Aytonian. It was strange that Bloyd's ship should be
called the *Three Sisters*; they must certainly have remarked on
this when they met in the *Grenville's* cabin, where Cook invited
him for a glass of wine.

In August 1766 Cook was at Cape Race and he observed an
eclipse of the sun. The details of his observation were later
published in the *Transactions* of the Royal Society and this
shows us, if we had not already realised it, that Cook was much
more than simply a competent navigator and surveyor. He had
the tireless curiosity so typical of the scientific mind. He also
had the care and exactitude so essential to scientific work. There
is no doubt he would have made a first rate astronomer. He
had the excellent eyesight, the capacity for accurate observation
and the manipulative skills essential to the mastery of the
instrumental aids of the working astronomer; his mathematical
abilities, too, would have been equal to the task. Indeed, he
was actually employed as the official astronomer during his
final voyage of discovery.

The planet Venus was calculated to pass over the sun's disc
in the middle of 1769. The Royal Society determined to send
observers here and there over the globe in order to observe this
rare occurrence, which happens only every hundred and twenty
years or so. Alexander Dalrymple was the man they proposed to
voyage into the southern hemisphere; or perhaps it might be
more correct to say that Dalrymple was the man who proposed
a voyage on behalf of the Society. But he had his conditions,
he wanted to be in charge of the ship as well as the expedition.
The Navy insisted that any ship of theirs must have a naval man
in command. Dalrymple sulked and resigned, which was just
as well for Mr Cook, Surveyor of Newfoundland. Whom better
to send? He was not only a naval man, but he had already
proved his scientific potential in his eclipse observations.

Guided by Palliser, the Admiralty appointed him to the
command of the *Endeavour Bark*, the ship selected for the
expedition.

CHAPTER 5

THE VOYAGE
OF THE ENDEAVOUR

'We had every advantage we could desire in
observing the whole of the passage of the planet
Venus over the sun's disc.'

The *Endeavour* could truly be said to be a Whitby ship, for
she was built there in the Fishburn yard and belonged to
Thomas Milner, who had named her the *Earl of Pembroke.*
She was chosen by Sir Hugh Palliser and Cook and none knew
better than he the capabilities of a flat bottomed Whitby collier.
It was large enough to hold all the stores for a long voyage, —
for they expected to be away at least two years — and at the
same time it was small enough to be steered through unknown
passages and in shallow waters with less danger than a traditional
navy ship.

In his choice of the Whitby cat, Cook showed his mastery of
the practical details and his grasp of the essentials to success in
long sea voyages of exploration. This going straight to the
heart of a problem and leaving out unessentials is a hall mark
of pioneer minds. Such men as Cook see clearly where they are
going and what they are to do, and gather around themselves,
by practical endeavour and force of will, all that is necessary
for success. As such figures recede into the past, aspects of
their lives and work become obvious that were by no means so
to their contemporaries. Now, all the time we see the mind of
Cook directing and organising, the type of ship, the care of
seamen's health and hygiene, the meticulous control of navigation and observation, and always that almost uncanny awareness
of what lay ahead, or just round the next headland.

By a wave of the Admiralty's wand the Whitby cat, *Earl of
Pembroke,* was transformed, — if not overnight, certainly in a
very short time — into His Majesty's Bark *Endeavour.* The
transformation might have been completed sooner, but strikes
held up the work. 'Tumults and riots of seamen' were not the
only source of trouble; the coal heavers and the Spitalfields
weavers were also protesting at the high cost of food and the low
wages.

Doubtless those on the *Endeavour* were glad to be sailing away from these troubles for a while. At least the crew were sure of their daily bread and there were enticing tales told by the men who had but just returned on Wallis's ship, the *Dolphin*. He had discovered some enchanted isles where the dusky, flower-decked maidens were all that a man dreamed of. It was on these very islands that the expedition proposed to observe the Transit of Venus. Could a more appropriately named planet have been chosen?

Tahiti was decided upon as the best place to make the observation on the advice of Captain Wallis. Except for the discovery of Tahiti and several islands in the Pacific, he had had a fruitless voyage in search of the mysterious continent which was supposed to exist somewhere in the southern hemisphere. The basis for the belief in the continent sounds strange to-day. The sages, Dalrymple among them, reasoned that there had to be an equally large lump of land south of the equator in order to balance all that which lay north of it. Otherwise the Earth would not be properly balanced. Yes indeed, there must be land down there somewhere, said the armchair geographers. Cook's instructions from the Admiralty said that, after observing the Transit, he was to go and search further for this elusive continent.

The work of preparation which went on is neatly summed up in Cook's *Journal*, which starts in May, 1768: 'At 11 am hoisted the pendant and took charge of the ship agreeable to my commission on the 25th instant. We lying in the basin at Deptford Yard. From this day to the 21st July we were constantly employed in fitting the ship, taking on board stores and provisions &c, when we sailed from Deptford and anchored in Galleons Reach.'

In August they sailed to Plymouth and on the 19th Cook 'Read to the ship's company the Articles of War and the Act of Parliament, they likewise were paid two month's wages in advance. I also told them not to expect any additional pay for the performance of our intended voyage. They were well satisfied and expressed great cheerfulness and readiness to prosecute the voyage.'

On Friday, 26th August 1768 'at 2 pm got under sail, and put to sea having on board 94 persons including Officers, Seamen, Gentlemen and their servants, near 18 months provisions, 10 carriage guns, 12 swivels with good store of ammunition and stores of all kinds.' And, one might add, a goodly stock of animals, not least a goat which had already sailed with Wallis.

The principal person on board after Cook was Joseph Banks, wealthy Lincolnshire land owner and botanist. He had the somewhat unusual gift, – for a man of his time and social position – of 'getting on' with almost anybody; even as a boy he made friends of the country herb-gathering women, paid them to collect plants and learnt much practical botany from them. This capacity for attaining intellectual ends through persons and events remained Banks's outstanding gift throughout his life. His family had risen to considerable affluence through good land management and progressive farming and Banks brought these family gifts to bear during his life-long dedication to British science.

To read the journal kept by Captain Cook is to read his mind and be with him on those momentous voyages. One is transported back two hundred years and one travels with him. Much has been written about his voyages, but it is mainly about the times spent on shore. Reading the day to day journal, one becomes enwrapt in the shipboard life. Day after day the wind direction is given:– 'Winds SSE to SE'; 'Winds E to ESE'; 'Winds NNE, NBW, SSW to SBW'. The distance travelled in miles is given:– 'Distce in miles 128', 'Distce in miles 79' or 140 or 68; the amount varies and is related to the wind direction.

Reading his journal, one all but feels the cold of the sea wind whipping around the deck of that sturdy little ship, day after day, night after night, riding the waves. 'First and latter part squally with heavy showers of rain,' it says, or: 'Fresh breeze and pleasant weather,' or: 'Moderate breeze and fair but cloudy and hazy over the land.'

One also gets the impression that Cook never slept at all. He seems to be on watch the whole time. This is partly due to the excellence of his officers who, reflecting their captain, make the observations as faithfully as he does himself. It was also partly an extra something which he possessed. After Cook's death a German sailor, Zimmerman, who went on the third voyage wrote:

> Fearlessness was his chief characteristic. He would
> run under full sail along the unknown coast of

America on foggy nights and sleep quietly through it
all; often, on the other hand, when no one suspected
any danger, he would come on deck and change the
course of the ship because land was nearer. So
pronounced was this faculty that everyone believed
he could instinctively sense and avert danger when it
was imminent. At any rate, I can swear that such
circumstances often arose when he alone, contrary to
all expectations, would become aware of the proxim-
ity of land, and his perception was always right.

Another fact which emerges from the *Journal* is that James
Cook was undoubtedly the captain and sole decision-maker on
that ship. 'I now determined to put into Rio de Janeiro in
preference to any other port,' we read, or: 'In this chart I have
laid down no land nor figured out any shore but what I saw
myself.'

He had an intuitive awareness of the vast pattern of sea and
sky that lay around and he sensed with amazing insight the lie
of land and sea. The years and years in Newfoundland had
sharpened this perception. In the latter part of this first voyage,
when he was delineating the coast line of New Zealand, his
air of sole authority and certain knowledge is almost uncanny.
'This point of land I have called East Cape, because I have
great reason to think that it is the easternmost land on this
whole coast,' . . . and it was. The same thing happened at
North Cape; he thought it must be, and it was. Later on he was
convinced that the northern part of New Zealand actually was
an island, but his officers were not so sure. He did not bother
to argue the point. He sailed round and through what is now
Cook Strait to the point the ship had left four months before.
He then 'called the officers upon deck and asked them if they
were now satisfied that this land was an island, to which they
answered in the affirmative, and we hauled our wind to the
Eastward.'

Cook's skilled use of Maskelyne's method of determining
longitude, which he had from the innovator himself – via his
assistant, Charles Green, who sailed on the *Endeavour* as
official Observer, – shows a rare aptitude and a capacity for
rapid assimilation of technical data.

The method, which was a considerable advance on any
previous practice involved reference to predicted lunar distances

for the prime meridian (Greenwich) which were supplied by the
new *Nautical Almanac,* first issued under Nevil Maskelyne's
direction in 1767. It consisted in observing the angular distance
of the moon from a fixed star, such an angle varying according
to the point on the Earth from which it was taken. It involved
more calculations than did the use of the chronometer, but
Cook was able by its means to record his discoveries with less
than one degree of longitudinal inaccuracy.

At the time of Cook's first voyage it was not generally
admitted that the Harrison chronometer had solved the
longitude problem and Cook sailed without one. In any case,
only one or two existed. Cook's use of Maskelyne's ingenious
innovation and his careful recording of his discoveries by its
means are yet another example of his taking to himself,
without hesitation, all that was needful for success in any enterprise, — the true mark of the man who, within any given
historical situation, stands head and shoulders above those
around him.

'We seldom failed of getting an observation every day to
correct our latitude by, and the observations for setting the
longitude were no less numerous and made as often as the sun
and moon came into play, so that it was impossible for any
material error to creep into our reckoning in the intermediate
time. In justice to Mr Green I must say that he was indefatigable
in making and calculating these observations, which otherwise
must have taken up a great deal of my time, which I could not
at all times very well spare. Not only this, but by his instructions
several of the petty officers can make and calculate these
observations almost as well as himself. It is only by such means
that this method of finding the longitude at sea can be put into
universal practice — a method that we have generally found may
be depended on to half a degree; which is a degree of accuracy
more than sufficient for all nautical purposes.'

'Would sea officers once apply themselves to the making and
calculating these observations, they would not find them so
very difficult as they first imagine, especially with the assistance
of the Nautical Almanac.'

From Plymouth the *Endeavour* steered for Madeira. Here,
the first punishments are recorded. Two seamen had twelve
lashes each for 'refusing to take their allowance of fresh beef.'
The word 'fresh' is questionable, for although the previous day
had been occupied in taking 'on board fresh beef and greens for

the ship's company', it is probable that any of the meat brought from England which remained would be served out before the new stores were touched. One can imagine just how 'fresh' this might be after three weeks in August, the height of the summer. Well did Cook know the ways of seamen who, seeing the newly shipped beef would refuse to eat the maggoty stuff served out to them. Right from the start he made sure they knew who was the captain and that his orders must be obeyed. The following day, however, he 'issued to the whole ship's company 20 pounds of onions per man.' After Madeira the diet was varied with fish for he 'served hooks and lines to the ship's company.'

Cook showed great subtlety in the matter of the sour krout, 'cabbage cut small and cured by going through a state of fermentation. It is afterwards close packed in casks with its own liquor, in which state it will keep any length of time. It is a very wholesome food and a very great anti-scorbutic.' At first the men would not eat it 'until I put into practice a method I never once knew to fail with seamen, and this was to have some of it dressed every day for the cabin table and permitted all the officers without exception to make use of it, and left it to the option of the men either to take as much as they pleased or none at all; but this practice was not continued above a week before I found it necessary to put everyone on board to an allowance, for such are the tempers and dispositions of seamen in general that whatever you give them out of the common way, although it be ever so much for their own good, yet it will not go down with them and you will hear nothing but murmurings against the man that first invented it. BUT the moment they see their superiors set a value on it, it becomes the finest stuff in the world and the inventor an honest fellow.'

Until this time scurvy had been the biggest single cause of death at sea. In 1753 a *Treatise of the Scurvy* was published by James Lind (1716–1794), a Scottish doctor who had himself served in the navy and seen its consequences first hand. In fact he had used some of his patients to make controlled experiments to find the cure.

Lind came from a large and substantial Edinburgh family, having several distinguished members, one of whom became Provost of the city and Member of Parliament. In 1758 James Lind was appointed chief physician of the Royal Naval Hospital at Haslar, near Portsmouth, where he served for twenty-five years. His cousin, also called James, became physician to the

JAMES LIND F.R.C.S. of Edinburgh. From a portrait painted after his appointment to Haslar. The Royal Naval Hospital can be seen in the background.

Royal Household at Windsor.

Lind dedicated the treatise to Lord Anson, who set off on his famous voyage with two thousand men and returned with two hundred, such were the ravages of scurvy. He made an intensive study of all the work which had gone before and found that 'what was first intended as a short paper swelled to a volume.' He found that he was obliged to discard some of the theories (It is caused by 'steams arising from the ocean'; 'corruption' and 'putrefaction'. Treatments including giving vitriol and bleeding — 'the patient usually dies'). In apologising to the 'eminent and learned authors' he says he has not 'done it with a malignant view of depreciating their labours, but from regard to truth and to the good of mankind.'

There is one author whose paper Lind quotes who had had the same regard for truth. This was Johann Friedrich Bachstrom, a Polish Lutheran pastor and medical doctor, who wrote a great many papers on many subjects well in advance of his time. One of these papers was on the subject of scurvy. It was written in Latin and the credit for its translation must go to Lind, but the

credit for the correct diagnosis of the cause and cure of scurvy
ought really to have been given to Bachstrom.

He it was who pointed out that scurvy was a land disease and
by no means peculiar to the sea alone. He had observed that it
frequently occurred in besieged towns, and in one succinct
sentence he put his finger on the crux:

> This evil is solely owing to a total abstinence from
> fresh vegetable food and greens, which is alone the
> true primary cause of the disease.

Bachstrom criticises the ineffectual use of drugs: 'The most
common herbs and fresh fruits excel the most pompous
pharmaceutical preparations.' This is the sort of remark one
comes across in studying the writings of original minds. The
freshness and validity of the insight cause the years to slip away
and bring the writer to one's elbow.

Among other things, Lind noted that the Newfoundland
people – whom Cook knew so well – kept free from scurvy
although they had no fresh vegetables during the winter. He
ascribed this to the spruce beer which they drank. Cook never
missed an opportunity to make this beverage wherever they
found pine trees. The use of wine also played its part and
accounts for the fact of naval officers being less susceptible to
scurvy than the sailors.

The day before Cook's fortieth birthday they 'got an observation and it was no longer doubted that we were southward of
the Line, the ceremony practised on this occasion by all nations
was not omitted. Everyone that could not prove upon a sea
chart that he had before crossed the Line was either to pay a
bottle of rum or be ducked in the sea, which former case was
the fate of by far the greatest part on board, and as several of
the men chose to be ducked and the weather was favourable for
that purpose, this ceremony was performed on about 20 or 30,
to the no small diversion of the rest.'

Cook's was by no means the only journal on board. The
twenty-five year old Joseph Banks, whose immense wealth had
played a large part in financing the voyage, kept a detailed and
lively account. On this occasion he purchased immunity for
himself, his party of fellow-naturalists, artists and servants, – to
say nothing of his two dogs – in brandy, but, he tells us, 'Many
of the men chose to be ducked rather than give up four days'

allowance of wine, which was the price fixed upon, & as for the boys they are always ducked of course, so that about 21 underwent the ceremony.'

There were boys a plenty aboard the *Endeavour*. One was Nicholas Young, whose name was destined to go on the map of the world, for he it was who first sighted land in New Zealand, and that point, *Young Nicks Head* remains to this day. Two boys whose names appear on the ship's company as officers' servants were not, however, actually present. They were James and Nathaniel Cook, aged 5 and 4 respectively, and their father entered their names according to naval custom, much as parents today enter infants for British public schools. The practice, though common, was actually forbidden by the Navy, but it was done to ensure a sufficiently long period of service for the young man in question so that promotion might be quicker. As it happened, both boys did enter the Navy, so it was as well their father had entered their names. Cook did not then know whether his fourth child, born after he left England, had been a boy or a girl, so he could not enter that name as well. In the event, the little boy, Joseph, lived only a month.

Several of the crew had been with Cook for some time, one of whom, Peter Flower, was unfortunately drowned at Rio de Janeiro. Cook noted, 'he was a good hardy seaman and had sailed with me above five years.'

The reception of the *Endeavour* at Rio de Janeiro was far from hospitable. The limitations of the Viceroy's astronomical knowledge was only matched by his suspicious and obtuse pugnacity. He did not believe that the *Endeavour* was a British Naval vessel and, says Cook, 'did not believe a word about our being bound to the southward to observe the transit of Venus, but looked upon it only as an invented story to cover some other design we must be upon, for he could form no other idea of that phenomenon (after I had explained it to him) than the North Star passing through the South Pole, (these were his own words).'

A guard was put on the ship's boats and restrictions placed on landing. Quaker ways and turns of speech in the Walker household had been unconsciously absorbed for here we find him saying: 'I was obliged to submit . . . being willing as much as in me lay to avoid all manner of disputes,' and again the following day: 'I was very loathe to enter into disputes.' Nevertheless, Cook got his own back for the Viceroy's hostile reception by making a detailed description of the bay and city of Rio de Janeiro, carefully noting all the fortifications in and around it, the number of troops and details of the government.

By the beginning of January 1769, the *Endeavour* was nearing Tierra del Fuego. They 'saw some penguins' and Cook 'gave to each of the people a Fearnought jacket and a pair of trousers, after which I never heard one man complain of cold, not but the weather was cold enough.'

By April, 1769, Cook had rounded the Horn and safely brought the little world contained in the *Endeavour* to Tahiti. When they went on shore 'no one of the natives made the least opposition at our landing, but came to us with all imaginable marks of friendship and submission.' The next day they had 'a great many canoes about the ship . . . It was a hard matter to keep them out of the ship for they climb like monkeys, but it was still harder to keep them from stealing but everything within their reach. In this they are prodigious expert. They are thieves to a man and would steal but everything that came their way, and that with such dexterity as would shame the most noted pickpocket in Europe.'

Cook had good reason to know of their thieving ability, for one night when he happened to sleep ashore, along with three other men, 'notwithstanding all the care we took, before 12 o'clock the most of us had lost something or other. For my own part I had my stockings taken from under my head, and yet I am certain that I was not asleep the whole time.'

On the whole the natives were a very clean race, though there was one thing which Cook noted 'that is disagreeable to Europeans, which is eating lice, a pretty good stock of which they generally carry about them.'

On the 17th April they were saddened by the death of one of their number. Cook records it thus: 'At 2 o'clock this morning departed this life Mr Alex Buchan, landscript draughts-- man to Mr Banks, a gentleman well skilled in his profession and one that will be greatly missed in the course of this voyage.' Banks's *Journal* records the event: 'His loss to me is irretrievable. My airy dreams of entertaining my friends in England with the scenes that I am to see here are vanished. No account of the figures and dress of men can be satisfactory unless illustrated with figures. Had Providence spared him a month longer what an advantage it would have been to my undertaking, but I must submit.'

In preparation for the observation of the Transit, Cook 'took as many people out of the ship as could possibly be spared and set about erecting a fort. Some were employed in throwing up entrenchments while others were cutting faccines, pickets &c. The natives were so far from hindering us that several assisted in bringing the pickets and faccines out of the woods and seemed quite unconcerned at what we were about. The wood we made use of for this occasion we purchased off them and we cut down no tree before we had first obtained their consent.'

At the very start of their stay in Tahiti Cook 'thought it very necessary that some order should be observed in trafficking with the natives; that such merchandise as we had on board for that purpose might continue to bear a proper value, and not leave it to everyone's particular fancy, which could not fail to bring on confusion and quarrels between us, and would infallibly lessen the value of such articles as we had to traffic with.' He therefore drew up a set of trading rules, one of which was that no sort of iron or anything made of iron was to be given for anything but provisions. This was an essential rule for 'they set great value on spike nails, but as this was an article many in the ship were provided with, the women soon found a much easier way of coming at them than by bringing provisions.'

George Young, writing from the standpoint of civilised virtue, tells us: 'They were so lascivious as to have no sense of modesty or decency. The females would openly lay themselves down and invite the strangers to their embraces. On one occasion a youth and a girl performed the rites of Venus in presence of a large company. At another time a female of rank paid her respects to Mr Banks by lifting her garments up to her waist, turning round three times and then dropping them.'

'It is much to be regretted,' says Young, 'that instead of discountenancing the licentiousness of the natives, the seamen, like those of the *Dolphin* and the two French ships commanded by Bougainville, disgraced themselves by indulging in illicit intercourse with the females of Tahiti. Such practices are not only highly immoral and degrading, but tend to the subversion of discipline and good order.'

Whatever Young may have thought about it, the seamen enjoyed themselves. 'Chastity is of but little value,' wrote Cook. 'The men will very readily offer the young women to strangers, even their own daughters, and think it very strange if you refuse them.' Cook recommended abstinence in order to stop the spread of venereal disease but, 'all I could do was to little purpose, for I may safely say I was not assisted by any one person in the ship.'

Nor was it only in this matter that Cook's was a lone voice, for with regard to the stealing, about which the natives were incorrigible, he says: 'I was very much displeased with them as they were daily either committing or attempting to commit one theft or another, when at the same time (contrary to the opinion of everybody) I would not suffer them to be fired upon.'

While the fort was being built they were visited by a great number of natives, among whom was Obarea, said to be queen of the island. She was 'about 40 years of age, and like most of the other women, very masculine. She is head or chief of her own family or tribe, but to all appearance hath no authority over the rest of the other inhabitants, whatever she might have had when the *Dolphin* was here.'

She brought presents 'which consisted of a hog, a dog, some breadfruits and plantains*. We refused to accept of the dog as being an animal we had no use for, at which she seemed a little surprised and told us it was very good eating, and we soon had an opportunity to find that it was so, for Mr Banks having bought a basket of fruit in which happened to be the thigh of a dog ready dressed, of this several of us tasted and found that it was meat not to be despised and therefore took Obarea's dog and had him immediately dressed in the following manner. They first made a hole in the ground in which they made a fire and heated some small stones. While this was doing the dog was strangled and the hair got off. His entrails were got out and the whole washed clean and as soon as the stones and hole were sufficiently heated the fire was put out and part of the stones left in the bottom of the hole, upon these stones were laid green leaves, and upon them the dog, together with the entrails. These were likewise covered with green leaves and over them hot stones and then the whole was close covered with mould. After he had lain there about four hours the oven was opened and the dog taken out whole and well done, and it was the opinion of everyone who tasted it that they never ate sweeter meat. We therefore resolved for the future not to despise dog's flesh.'

*Plantains: ie: Bananas.

The great day of the observation of the transit of Venus dawned with a clear sky, 'not a cloud was to be seen the whole day and the air was perfectly clear so that we had every advantage we could desire in observing the whole of the passage of the planet Venus over the sun's disc. We very distinctly saw an atmosphere or dusky shade round the body of the planet which very much disturbed the times of the contacts.' Cook and Green, the astronomer sent out by the Royal Society, made careful observations of the transit at Fort Venus, while two other parties at considerable distances on either side made independent observations. The atmosphere which Cook saw and mentions made marked differences in the determination of the transit and in this expedition, – as in much later ones (1874 and 1882) – led to failure. In fact, the transit of Venus is not a method capable of yielding a good solution of the sun's distance from the Earth, which has been solved by other means. The minor planets, or asteroids, having no atmosphere afford much more accurate data.

Later, when the results were given to Maskelyne he said they differed 'more from one another than they ought to do and cannot account for it in no other way than the want of care and address in the observers. Mr M. might have assigned another reason,' wrote Cook. 'He was not unacquainted with the quadrant having been in the hands of the natives, pulled to pieces and many of the parts broke which we had to mend in the best manner we could before it could be made use of. Mr M. should have considered before he took it upon himself to censure these observations, that he had put into his hands the very original book in which they were written in pencil only, the very moment they were taken. . . . Mr M. should also have considered that this was perhaps the only true, original paper of the kind ever put into his hands. Does Mr M. publish to the world all the observations he makes, good and bad, or did he never make a bad observation in his life?'

Their purpose for visiting Tahiti being accomplished, they prepared to leave and on 12th July 1769: 'The carpenters employed stocking the anchors and the seamen in getting the ship ready for sea.'

For some time before they left 'several of the natives were daily offering themselves to go away with us, and as it was thought they must be of use to us in our future discoveries, we resolved to bring away one whose name is Tupia.' Cook himself was a little doubtful but, 'at the request of Mr Banks I received him on board, together with a young boy, Tiato, his servant.' Banks's own journal gives an interesting account of the reasons for his request: 'I do not know why I may not keep him as a curiosity, as well as some of my neighbours do lions and tigers, and at a larger expense than he will probably ever put me to. The amusement I shall have in his future conversation and the benefit he will be to this ship . . . will, I think, fully repay me.'

In the next year and a half of the voyage Tupia did indeed prove an invaluable aid. He was able to act as interpreter and guide right through the Pacific Islands and even as far as New Zealand. Neither he nor the boy ever reached England, however, for in Batavia they 'fell a sacrifice to this unwholesome climate before they had reached the object of their wishes.' Indeed, Tupia's death was not wholly owing to the 'unwholesome air of Batavia. The long want of a vegetable diet which he had all his life before been used to, had brought upon him all the disorders attending a sea life. He was a shrewd, sensible, ingenious man,' Cook wrote after his death, 'but proud and obstinate, which often made his situation on board both disagreeable to himself and those about him, and tended much to promote the diseases which put a period to his life.'

CHAPTER 6

THE GREAT BARRIER REEF

'Such are the vicissitudes attending this kind of
service and must always attend an unknown navigation. Was it not for the pleasure which naturally
results to a man from being the first discoverer . . .
this service would be insupportable.'

Leaving Tahiti, Cook took the Endeavour to four nearby islands
which 'we were informed lay only one or two days' sail to the
westward of Georges Island, and that we might there procure
hogs, fowls and other refreshments, articles that we had been
very sparely supplied with at this last island; and as the ship's
company, what from the constant hard duty they had had at
this place and the too free use of women, were in a worse state
of health than they were on our first arrival; for by this time
full half of them had got the venereal disease, in which situation
I thought they would be ill able to stand the cold weather we
might expect to meet with to the southward at this season of
the year, and therefore I resolved to give them a little time to
recover while we run down to and explored the islands before
mentioned.'

At Huahine, Cook, Surveyor of Newfoundland, 'set about
surveying the island, and Dr Munkhouse with some hands went
ashore to trade with the natives, while the long boat was
employed completing our water.'

'The produce of this island,' wrote Cook, 'is in all respects
the same as King Georges Island, and the manners and customs
of the inhabitants much the same, only that they are not
addicted to stealing and with respect to colour they are rather
fairer.'

At Otaha he 'sent the Master in the long boat with orders to
sound the harbour, and if the wind did not shift in our favour to
land upon the island and to traffic with the natives for such
refreshments as were to be got. Mr Banks and Dr Solander went
along with him.' But they stayed away longer than Cook wished
so he 'fired a gun for her to return and as soon as it was dark

hoisted a light.'

After this he 'steered to the southward' and on 14th August, 1769, they reached the island called Rurutu but 'found that there was neither a harbour or safe anchorage about it and therefore I thought the landing upon it would be attended with no advantage either to ourselves or any future navigators, and from the hostile and thievish disposition of the natives it appeared we could have no friendly intercourse with them until they had felt the smart of our firearms, a thing that would have been very unjustifiable in me at this time, we therefore hoisted in the boat and made sail to the southward . . . being now fully resolved to stand directly to the southward in search of the Continent.'

Today, looking at the route which Cook took, one has a strange impression that he knew where he was going. Perhaps he did. Apart from one small circle (occasioned by contrary winds) the line is practically a direct one from the Society Islands to New Zealand. Seen on a chart, it could easily be taken for the course of a modern ship with all the advantages of power, gyro compasses, radar and radio. Cook had only the wind in his sails and Maskelyne in his mind.

After nearly two month's sailing, hints of land began to appear. 'At noon saw some seaweed.' A few days later: 'Saw an immense number of birds, the most of them were doves. Saw likewise a seal asleep upon the water which we at first took for a crooked billet. These creatures as they lay upon the water hold their fins up in a very odd manner and very different to any I have seen before.' On the 5th October: 'Saw a great many porpoises, large and small. Saw two Port Egmont hens, a seal, some seaweed and a piece of wood with barnacles upon it.' Then on Saturday the 7th of October 1769, 'At 2 pm saw land from the mast head.' The one who saw it was a lad, Nicholas Young and the headland he spotted Cook named *Young Nick's Head*.

Though Tasman had sailed here, he had never actually landed. Very soon Cook 'went ashore with a party of men in the pinnace and yawl accompanied by Mr Banks and Dr Solander.' Not surprisingly, the natives were hostile although 'Tupia spoke to them in his own language and it was an agreeable surprise to us to find that they perfectly understood him.' Satisfactory contact was not made until the *Endeavour* had fired on and even killed some of them. Perhaps Cook was thinking of the Quaker brothers at Whitby when he wrote: 'I am aware that

most humane men will censure my conduct in firing upon the
people in this boat, nor do I think myself that the reason I had
for seizing upon her will at all justify me, and had I thought
that they would have made the least resistance I would not have
come near them, but as they did I was not to stand still and
suffer either myself or those that were with me to be knocked
on the head.'

A day or two later Cook left the 'bay which I have named
Poverty Bay because it afforded us no one thing we wanted.'
Running into shoal water 'four canoes came off to us full of
people and kept for some time under our stern, threatening us
all the while. As I did not know but what I might be obliged
to send our boats ahead to sound, I thought these gentry would
be as well out of the way, so I ordered a musket shot to be fired
close to one of them.'

Sailing south, 'some fishing boats came off to us and sold us
some stinking fish. One man in this boat had on him a black
skin, something like a bearskin, which I was desirous of having
that I might be a better judge of what sort of animal the first
owner was.' But instead of handing over the skin in return for
the things they were given, they seized Tiato and 'endeavoured
to carry him off. This obliged us to fire upon them, which gave
the boy an opportunity to jump overboard and we took him
up unhurt. This affair occasioned my giving this point of land
the name of *Cape Kidnappers.'*

In the ensuing six months the *Endeavour* sailed right around
the two islands of New Zealand. The voyage was not without
incident. They found that some of the natives were friendly
and ready to trade with them, while others were hostile. Many
of them were undoubtedly cannibals.

The journal contains many references to the use of firearms.
At one place when they went ashore 'before we could well look
about us we were surrounded by 2 or 3 hundred people, and
notwithstanding that they were all armed, they came upon us
in such a confused, straggling manner that we hardly suspected
that they meant us any harm, but in this we were very soon
undeceived, for upon us endeavouring to draw a line upon the
sand between us and them, they set up the war dance and
immediately some of them attempted to seize the two boats.'
Muskets loaded with small shot were fired and the ship fired
over their heads. 'In this skirmish only one or two of them was
hurt with small shot, for I avoided killing any one of them as

much as possible, and for that reason withheld our people from firing.'

On another occasion, when Cook 'gave everybody leave to go ashore at the watering place to amuse themselves as they thought proper' the master and 'five petty officers desired to have a small boat to go a fishing.' Some of the natives 'put off in two canoes, as they thought to attack them. This caused the master to fire and wounded two, one of which is since dead . . . I find the reason for firing upon them not very justifiable,' wrote Cook.

In November, 1769, the transit of another planet, Mercury, was observed by Charles Green, the Royal Society's official observer. Cook 'at this time was taking the sun's altitude in order to ascertain the time.' While he was thus occupied a canoe came off to the ship and Lieutenant Gore had an altercation with a native over a piece of cloth which he bartered. 'Mr Gore* fired a musket at them, and from what I can learn, killed the man who took the cloth, and after this they soon went away. I have here inserted the account of this affair just as I had it from Mr Gore, but I must own that it did not meet with my approbation because I thought the punishment a little too severe for the crime, and we had now been long enough acquainted with these people to know how to chastise trifling faults like this without taking away their lives.'

All was not by any means plain sailing with the naturalists aboard ship. There were several recorded occasions when Cook differed in opinion, but he was the captain and had the last word, however disgruntled they may have been. For example, at West Cape they wanted to go right in to a narrow place to harbour, but Cook 'saw clearly that no winds could blow there but what was either right in or right out. This is Westerly or Easterly and it certainly would have been highly imprudent in me to have put into a place where we could not have got out but with a wind which we have lately found does not blow one day in a month. I mention this because there were some on board who wanted me to harbour at any rate without in the least considering either the present or the future consequences.'

By 27th March 1770 Cook could say: 'As we have now circumnavigated the whole of this country, it is time for me to

---

*Lieutenant John Gore, an American. Later to bring home the remnants of the final expedition.

think of quitting it, but before I do this it will be necessary first
to complete our water.' Accordingly, the careful Cook, who
might well have sent an officer to do it, himself 'took a boat
and went to look for a watering place and a proper berth to
moor the ship in, both of which I found convenient enough.'

On Sunday, 1st April 1770 they finally left New Zealand
'with an intention to steer to the westward, which we accordingly
did, taking our departure from Cape Farewell.' By the 17th they
were nearing Australia and 'a small land bird was seen to perch
upon the rigging' and a couple of days later they 'saw land
extending from NE to West at the distance of 5 or 6 leagues,
having 80 fathom water, a fine sandy bottom.' Cook named the
point of land 'Point Hicks because Lieut. Hicks was the first
who discovered this land.'

The first attempt to land was unsuccessful for 'we no where
could effect a landing by reason of the great surf which beat
everywhere upon the shore.' But the following day, Sunday the
29th April 1770, they did land. Cook, about to step ashore, was
suddenly impressed by the import of this momentous occasion
and turning to his wife's young cousin, Isaac Smith, said: 'Isaac,
you shall land first'. The boy remembered this action all the
rest of his long life.*

They found the natives extremely shy and in a far more
primitive state than the others they had already met, 'they were
quite naked, even the woman had nothing to cover her nudity.'

Towards the end of May, 1770 a 'very extraordinary thing
happened to Mr Orton, Cook's clerk, 'he, having been drinking
in the evening, some malicious person or persons in the ship
took the advantage of his being drunk and cut off all the clothes
from off his back. Not being satisfied with this, they went into
his cabin and cut off a part of both his ears as he lay asleep in
his bed. The person whom he suspected to have done this was
Mr Magra, one of the midshipmen. ... I therefore for the
present dismissed him from the quarter deck and suspended
him from doing any duty in the ship, he being one of those
gentlemen, frequently found on board King's ships, that can
well be spared, or to speak more plainly, good for
nothing.'

Steadily, they sailed up the east coast of Australia, sometimes
landing, sometimes not. Not all the points of land were named

*He eventually became an admiral and in later life resided with Mrs Cook.

after officers and distinguished persons. Cook records: 'Last night Torby Sutherland, seaman, departed this life and in the am his body was buried ashore at the watering place, which occasioned my calling the south point of this bay after his name.'

They were fast approaching the Great Barrier Reef, that very beautiful but deadly jagged wall of sharp rocky coral, for the most part submerged, which extends for hundreds of miles along the North easterly coast of Australia. On the 10th June, 1770, Cook passed a point which he later named *Cape Tribulation* 'because here began all our troubles.' Reading Cook's journal gives one a vivid sense of being there with him through all those terrifying weeks after 'the ship struck and stuck fast.'

We are with him when he throws overboard 'our guns, iron and stone ballast, casks, hoops, stays, oil jars, decayed stores &c.' We feel the anxiety when 'the leak gained upon the pumps considerably. This was an alarming and, I may say terrible circumstance, and threatened immediate destruction to us ... A mistake soon after happened which for the first time caused fear to operate upon every man in the ship. The man which attended the well took the depth of water above the ceiling; he being relieved by another who did not know in what manner the former had sounded, took the depth of water from the outside plank, the difference being 16 or 18 inches, and made it appear that the leak had gained this much upon the pumps in a short time. This mistake was no sooner cleared up than it acted upon every man like a charm: they redoubled their vigour in so much that before 8 o'clock in the morning they gained considerably upon the leak.'

It was decided to apply a fothering sail. Cook had no hesitation in appointing one of his midshipmen, Mr Munkhouse, to oversee this operation, for he 'was once in a merchant ship which sprung a leak, but by this means was brought home from Virginia to London with only her proper crew. To him I gave the direction of this, who executed it very much to my satisfaction.' The fothering sail was filled with oakum and wool loosely stuck with dung, then it was put beneath the ship and sucked into the hole, thus partly blocking it.

Cook then sent the 'Master with two boats as well to sound ahead of the ship as to look out for a harbour where we could repair our defects and put the ship into a proper trim.' The weather began to blow hard, so Cook anchored and 'made the

signal for the boats to come on board, after which I went myself
and buoyed the channel, which I found very narrow and the
harbour much smaller than I had been told, but very convenient
for our purpose.'

When they eventually got the ship ashore and 'the tide left
her which gave us an opportunity to examine the leak' they
found 'the rocks had made their way through four planks quite
to and even into the timbers and wounded three more. The
manner these planks were damaged, or cut out as I may say, is
hardly credible. Scarce a splinter was to be seen, but the whole
was cut away as if it had been done by the hands of a man with
a blunt edged tool.'

'The carpenters went to work on the ship, while the smiths
were busy making bolts, nails &c.' In four days, between tides,
they had finished. They did their work well, for a month after
Cook wrote: 'I cannot complain of a leaky ship, for the most
water she makes is not quite an inch an hour.'

While the *Endeavour* was being mended some of the crew were
sent to gather greens for the ship's company while others were
occupied in catching turtles. One entry in Cook's *Journal* made
us prick up our Whitby ears: 'At 2 o'clock the yawl came on
board and brought three turtle and a large skeat.' This is the
way that skate is pronounced in Whitby, — and Staithes too.
We wonder if Cook wrote the word as he had heard it pronounced there.

'Whatever refreshment we got that would bear a division I
caused to be equally divided amongst the whole company
generally by weight. The meanest person in the ship had an equal
share with myself or anyone on board, and this method every
commander of a ship on such a voyage as this ought ever to
observe.'

They managed to make contact with a few natives. 'About
this time 5 of the natives came over and stayed with us all the
forenoon. There were 7 in the whole, 5 men, a woman and a
boy. These two last stayed on the point of sand on the other side
of the river about 200 yards from us. We could very clearly see
with our glasses that the woman was as naked as ever she was
born, even those parts which I always before now thought Nature
would have taught a woman to conceal were uncovered.'

The natives 'that came on board were very desirous of having
some of our turtle and took the liberty to haul two to the
gangway to put over the side; being disappointed in this they

grew a little troublesome and were for throwing everything overboard they could lay their hands upon. As we had no victuals dressed I offered them some bread to eat, which they rejected with scorn, as I believe they would have done anything else except turtle. Soon after this they all went ashore, Mr Banks, myself and five or six of our people being ashore at the same time. Immediately upon their landing one of them took a handful of dry grass and lighted it at a fire we had ashore, and before we well knew what he was about he made a large circuit about us and set fire to the grass in this way. In an instant the whole place was in flames.' Not content with this, they 'went to a place where some of our people were washing and where all our nets and a good deal of linen were laid out to dry. Here, with the greatest obstinacy, they again set fire to the grass, which I and some others could not prevent, until I was obliged to fire a musket loaded with small shot at one of the ring leaders, which sent them off.'

Banks and Cook went up 'a high hill from whence we had an extensive view of the sea coast to leeward; which afforded us a melancholy prospect of the difficulties we are to encounter, for in whatever direction we turned our eye, shoals innumerable were to be seen.'

By 4th August, 1770, all was ready for departure and Cook was impatient to be off, for 'laying in port spends time to no purpose, consumes our provisions, of which we are very short in many articles, and we have yet a long passage to make to the East Indies through an unknown and perhaps dangerous sea; these circumstances considered make me very anxious of getting to sea.'

They put to sea but it was indeed 'a melancholy prospect.' Cook himself and 'several of the officers kept a look-out at the mast head to see for a passage between the shoals, but we could see nothing but breakers all the way, from the south round by the east as far as NW, extending out to sea as far as we could see.'

By the 14th August, 1770, Cook could write: 'We got safe out. We had no sooner got without the breakers than we had no ground with 150 fathom of line and found a well grown sea rolling in from the SE, certain signs that neither land nor shoals were in our neighbourhood in that direction, which made us quite easy at being freed from fears of shoals &c — after having been entangled in them more or less ever since the 26th of May, in which time we have sailed 360 leagues without ever having a man out of the chains heaving the lead when the ship was under way, a circumstance I daresay never happened to any ship before, and yet here it was absolutely necessary.'

'It was with great regret I was obliged to quit this coast unexplored to its northern extremity, for I firmly believe it doth not join to *New Guinea*, however, this I hope yet to clear up, being resolved to get in with the land again as soon as I can do it with safety.'

But they were not yet clear of the reef. Two days later they found they were fast approaching 'vast foaming breakers' and their depth of water was unfathomable so that there was not a possibility of anchoring. 'In this distressing situation we had nothing but Providence and the small assistance our boats could give us to trust to . . . We had hardly any hopes of saving the ship, and full as little our lives, as we were full ten leagues from the nearest land and our boats not sufficient to carry the whole of us, yet in this truly terrible situation not one man ceased to do his utmost, and that with as much calmness as if no danger had been near. All the dangers we had escaped were little in comparison of being thrown on this reef, where the ship must be dashed to pieces in a moment.'

'A reef such as is here spoken of is scarcely known in Europe. It is a wall of coral rock rising almost perpendicular out of the unfathomable ocean, always overflown at high water generally 7 or 8 feet and dry at places at low water. The large waves of the vast ocean meeting with so sudden a resistance make a most terrible surf, breaking mountains high.'

'At this critical juncture, when all our endeavours seemed too little, a small air of wind sprung up, but so small that at any other time in a calm we should not have observed it.' With its assistance and the turn of the tide they managed to navigate the ship into a sheltered part of the reef, 'happy once more to encounter those shoals which but two days ago our utmost wishes were crowned by getting clear of. So much does great danger swallow up lesser ones that those once so dreaded shoals were now looked at with less concern.'

'Such are the vicissitudes attending this kind of service and must always attend an unknown navigation. Was it not for the pleasure which naturally results to a man from being the first discoverer, even was it nothing more than sands and shoals, this service would be insupportable, especially in far distant parts like this, short of provisions and almost every other necessity.

The world will hardly admit of an excuse for a man leaving a coast unexplored he has once discovered. If dangers are his excuse then he is charged with *timorousness* and want of perseverance and at once pronounced the unfittest man in the world to be employed as a discoverer. If on the other hand he boldly encounters all the dangers and obstacles he meets and is unfortunate enough not to succeed, then he is charged with *temerity* and want of conduct.'

'I now came to a fixed resolution to keep the main land on board in our route to the northward, let the consequence be what it will.' So the *Endeavour* threaded her way right up to the northernmost tip of Australia and Cook saw 'an open sea to the westward, which gave me no small satisfaction, not only because the dangers and fatigues of the voyage were drawing near to an end, but by being able to prove that New Holland and New Guinea are two separate lands or islands, which until this day hath been a doubtful point with geographers.'

He thought that there might be a safer passage than the one he had taken, but 'this was a thing I had neither time nor inclination to go about, having been already sufficiently harassed with dangers without going to look for more.'

As they left Australia, where, said Cook, 'it can never be doubted but what most sorts of grain, fruits and roots &c of every kind would flourish were they once brought hither, planted and cultivated by the hand of industry,' they touched at New Guinea. Cook 'went ashore in the pinnace accompanied by Mr Banks and Dr Solander, having a mind to land once in this country before we quit it altogether, which I am now determined to do without delay.'

And so 'upon my return to the ship we hoisted in the boat and made sail to the westward with a design to leave the coast altogether, to the no small satisfaction of, I believe, the major part of ye ship's company. However, it was contrary to the inclination and opinion of some of the officers, who would have had me send a party of men ashore to cut down the cocoanut trees for the sake of the nuts, a thing that I think no man living could have justified; for as the natives had attacked us for mere landing without taking away any one thing, certainly they would have made a vigorous effort to have defended their property, in which case many of them must have been killed, and perhaps some of our own people too — and all of this for 2 or 300 green cocoanuts, which when we had got them would have done us little service. Besides, nothing but the utmost necessity would have obliged me to have taken this method to come at refreshments.'

At Batavia, after much intercourse with Dutch officialdom, the ship was 'delivered into the charge of the proper officers at Onrust, who will (I am informed) heave her down and repair her with their own people only, while ours must stand and look on.'

They found the ship's bottom 'to be in a far worse condition than we expected. The false keel was gone to within 20 feet of the stern post; the main keel wounded in many places very considerably; a great quantity of sheathing off, several planks much damaged . . . so that it was a matter of surprise to everyone who saw her bottom how we had kept her above water; and yet in this condition we had sailed some hundreds of leagues in as dangerous a navigation as in any part of the world, happy in being ignorant of the continual danger we were in.'

Perhaps it was as well that the Batavians insisted on their own men doing the work, for after a month; 'we were so weakened by sickness that we could not muster above 20 men and officers that were able to do duty.' They had come to Batavia 'with as healthy a ship's company as need go to sea, and after a stay of not quite three months, left it in the condition of a hospital ship, besides the loss of 7 men.'

Their troubles were by no means over. Even today, two hundred years after, it is distressing to read the pages of Cook's *Journal* of that time, so filled with irrevocable tragedy as they are. Leaving out the details of course and the state of the weather, these are the sort of entries we read:

> Thursday 24th January 1771. In the am died Jno Truslove, Corporal of Marines. A man much esteemed by everyone on board. Many of our people at this time lay dangerously ill of fevers and fluxes.

> Friday 25th. Departed this life Mr Sporing, a gentleman belonging to Mr Banks's retinue.

> Sunday 27th. Departed this life Mr Sidney Parkinson, natural history painter to Mr Banks, and soon after Jno Ravenhill, a man much advanced in years.

Tuesday 29th. In the night died Mr Charles Green, who was sent out by the Royal Society to observe the Transit of Venus; he had long been in a bad state of health, which he took no care to repair, but on the contrary lived in such a manner as greatly promoted the disorders he had had long upon him, this brought on the flux which put a period to his life.

Wednesday 30th. Died of the flux, Saml Moody and Francis Hate, two of the carpenter's crew.

Thursday 31st. In the course of this 24 hours we have had four men die of the flux, viz: Jno Thompson, ship's cook, Benj Jordan, carpenter's mate, James Nicholson and Archd Wolfe, seamen. A melancholy proof of the calamitous situation we are at present in, having hardly well men enough to tend the sails and look after the sick, many of the latter are so ill that we have not the least hopes of their recovery.

It is a depressing part of the journal and does not end until the 27th February, 1771:

In the am died of the flux Henry Jeffs, Emanuel Pharah and Peter Morgan, seamen. The last came sick on board at Batavia of which he never recovered and the other two had long been past all hopes of recovery, so that the death of these three men in one day did not in the least alarm us; on the contrary we are in hopes that they will be the last that will fall a sacrifice to this fatal disorder, for such as are now ill of it are in a fair way of recovering.

In the middle of March, 1771, they limped into Cape Town. Cook's 'first care was to provide a proper place ashore for the reception of the sick, for which purpose I ordered the Surgeon to look out for a house where they could be lodged and dieted.'
Comparing notes with other ships in Cape Town he found that 'ships which have been little more than twelve months from England have suffered as much or more by sickness than we have done who have been out near three times as long. Yet their sufferings will hardly, if at all, be mentioned or known in

England, when on the other hand those of the *Endeavour*,
because the voyage is uncommon, will very probably be
mentioned in every newspaper, and what is not unlikely, with
many additional hardships we never experienced. For such are
the dispositions of men in general in these voyages that they
are seldom content with the hardships and dangers which will
naturally occur, but they must add others which hardly ever
had existence but in their imaginations, by magnifying the
most trifling accidents and circumstances to the greatest
hardships and unsurmountable dangers, without the immediate
interposition of Providence, as if the whole merit of the voyage
consisted in the dangers and hardships they underwent; or that
real ones did not happen often enough to give the mind sufficient
anxiety. Thus posterity is taught to look upon these voyages as
hazardous to the highest degree.'

In April they sailed from Cape Town, the sick people being
mostly recovered, but it was a sadly depleted ship's company.
Green's absence is particularly noticeable in one entry which
Cook made: 'In the pm observed merely for the sake of observing
an Eclipse of the Sun.' Towards the end of May, 'departed this
life Lieut. Hicks and in the evening his body was committed to
the sea with the usual ceremonies. He died of a consumption
which he was not free from when we sailed from England, so
that it may be truly said that he had been dying ever since,
though he held out tolerably well until we got to Batavia.' In
Hicks's place Cook appointed Charles Clerke, 'he being a young
man extremely well qualified for that station.' This was the
same Clerke who ultimately took command after Cook's death,
but died of consumption before reaching England. We wonder
if he first contracted the disease in Hicks's cabin.

For a time the *Endeavour* sailed towards home in company
with a fleet of '13 sail of stout ships, which we took to be the
East India fleet.' Her rigging and sails were now so badly worn
that 'something or another is giving way every day.' On 7th
July, 1771 they 'spoke a Brig from Liverpool and some time
after another from London. . . . We learnt from this vessel that
no accounts had been received in England and that wagers were
held that we were lost.'

On 10th July, 1771 they 'saw land from the masthead bearing
north, which we judged to be about Lands End,' and on the 13th
they anchored in the Downs and soon after Cook 'landed in
order to repair to London.'

CHAPTER 7

THE SECOND VOYAGE

'The public must not expect from me the elegance
of a fine writer ... but will, I hope, consider me as a
plain man, zealously exerting himself in the service
of his country.'

Even before he had landed in England, Cook had written the
Postcript to his Journal. It concludes:

Now I am upon the subject of discoveries, I hope
it will not be taken amiss if I give it as my opinion
that the most feasible method of making further
discoveries in the South Sea is to enter it by way of
New Zealand, first touching and refreshing at the Cape
of Good Hope, from thence proceed to the southward
of New Holland for Queen Charlotte's Sound, where
again refresh wood and water, taking care to be ready
to leave that place by the latter end of September,
where you would have the whole summer before you,
and after getting through the strait might, with the
prevailing westerly winds run to the eastward in as
high a latitude as you please, and if you meet with
no lands would thus have time enough to get round
Cape Horn before the summer was too far spent, but
if after meeting with no continent, & you had other
objects in view, then haul to the northward ... thus
the discoveries in the South Sea would be complete.

It was obvious that after refreshment in England Cook wanted
to return to his discovering life. The Admiralty were not slow
in accepting his suggestion. He was asked to find suitable ships,
for this time it was decided to send a support vessel with him.
Their Instructions were drawn upon consultation with him and
embraced all that he had suggested in the *Postcript.*
Again Cook chose Whitby ships for (writing to John Walker
after his return) 'A better ship for such a service I never would

61

wish for.' The *Marquis of Granby* and the *Marquis of Rockingham* were the ships selected and they were eventually re-named *Resolution* and *Adventure.*

Cook stayed exactly one year in England. Most of the time was spent in preparing the ships for their intended voyage. His return to his family was a sad one, for their little daughter had died only a few months before and he also learnt that the baby whom he had never seen, little Joseph, had not survived more than a month. However, on this leave the balance was restored, for their fourth son, George – a royal name – appeared one month before his father sailed. Sadly, however, the baby did not last the winter out and he too followed Joseph.

Cook was promoted to the rank of Captain and he was honoured by being presented to the young King George III, 'who was pleased to express his approbation of my conduct in terms that were extremely pleasing to me.'

Although he had crossed the wide oceans to unknown parts, a journey from London to the North of England was no small undertaking, especially when he was accompanied by his wife, who was pregnant. It is recorded that in December 1771 he visited Ayton and Whitby. There may well have been other visits of which no record remains, but we must imagine the difficulties of eighteenth century travel and also remember how short were Cook's periods of leave, and all the claims of a young family. On this particular visit he had intended returning by sea from Hull, but 'Mrs Cook being but a bad traveller, I was prevailed upon to lay that route aside.' This must be one of the very few instances when Cook changed his mind and did not follow his own intended plan. It is a revealing instance of his devotion to his wife.

The following year his father left Ayton and went to live with his daughter, Margaret Fleck, – 'the wife of a respectable fisherman in Redcar', Young tells us. She had a large family and many of her descendants survive today and proudly trace back their link to this first James Cook. None of Captain Cook's own family had children, so there are no direct descendants.

While at Ayton, Cook met Commodore William Wilson, whose chief posthumous claim to fame was that he had been a friend of the celebrated circumnavigator. An elegant 18th century marble memorial to him in the old church at Ayton (in the graveyard of which lies Cook's mother) depicts in bas relief the East India ships in which he sailed and records his active maritime life. It does not, however, record what he once said, namely: 'The English have a right to navigate wherever it has pleased God to send water.'

Although the following *Preface* was not actually written until after Cook's return from his second voyage, we feel it right to quote it here, in the correct context, before he had actually left.

> And now it may be necessary to say that as I am on the point of sailing on a third expedition, I leave this account of my last voyage in the hands of some friends, who in my absence have kindly accepted the office of correcting the press for me; who are pleased to think that what I have here to relate is better to be given in my own words than in the words of another person; especially as it is a work designed for information and not merely for amusement; in which, it is their opinion that candour and fidelity will counterbalance the want of ornament.
>
> I shall therefore conclude this introductory discourse with desiring the reader to excuse the inaccuracies of style, which doubtless he will meet with in the following narrative; and that, when such occur he will recollect that it is the production of a man who has not had the advantage of much school education, but who has been constantly at sea from his youth; and though with the assistance of a few good friends, he has passed through all the stations belonging to a seaman, from an apprentice in the coal trade to a post captain in the Royal Navy, he has had no opportunity of cultivating letters.
>
> After this account of myself, the public must not expect from me the elegance of a fine writer or the plausibility of a professed book maker, but will, I hope consider me as a plain man, zealously exerting himself in the service of his country, and determined to give the best account he is able of his proceedings.

At first 'Mr Banks and Dr Solander who accompanied me in my last voyage intended to embark with me in this in order to prosecute their discoveries in natural history and botany and other useful knowledge' and 'the Parliament voted four thousand pounds towards carrying on discoveries to the South Pole, this sum was intended for Dr Lynd* of Edinburgh as an encouragement for him to embark with us, but what the discoveries were the Parliament meant he was to make, and for which they made so liberal a vote, I know not.' In order to accommodate the extra passengers 'the Navy Board was prevailed upon, – though contrary to the opinion of some of the members, particularly the comptroller – to alter their former plan, which was to leave her in her former state and to raise her upper works about a foot, to lay a spar deck upon her ... and to build a round house or couch for my accommodation so that the great cabin might be appropriated to the use of Mr Banks alone.'

Cook 'had reason to think that she would prove crank and that she was over built.' Sure enough, when she was put on her trials this proved to be the case. 'She was found to be so crank that it was thought unsafe to proceed any further with her.' The Admiralty ordered the superstructure to be removed, but this did not suit the young Banks. 'I shall not mention the arguments made use of by Mr Banks and his friends as many of them were highly absurd and advanced by people who were not judges of the subject,' wrote Cook, who had known all along how unsuitable the plans were, but had remained silent until the end.

Thus it was that 'Mr Banks declared his resolution not to go the voyage, alleging that the sloop was neither roomy nor convenient enough for his purpose, nor noways proper for the voyage.' To her captain, however, 'she was the ship of my choice ... and the properest ship for the service she is intended for of any I ever saw.'

The comptroller of the Navy whose advice had been over ruled was Cook's earliest friend and patron in the service, Hugh Palliser, He visited the ship after Banks and his suite, with their attendant baggage, had left it, 'for this gentleman had taken it upon him, in spite of all that had been alleged against her, to make her completely fit not only for the sea but for the service she was intended for. Indeed, if his advice had not been over ruled at first, a great deal of unnecessary trouble and expense would have been saved, not only to the Crown, but to Mr Banks and every other person concerned.'

*This Dr James Lind was a second cousin of the celebrated Lind of the *Treatise on Scurvy*. At the time he practised medicine in Edinburgh, but later went to Windsor, where he was physician to the Royal household. His cousin was in charge of the Royal Naval Hospital at Portsmouth at this time.

The Earl of Sandwich also 'anchored at the Nore and soon after landed in the yard and came on board the *Resolution.* His lordship inspected and was pleased to approve of all the alterations that had been made.'

A final visit to London where Cook 'learnt that Mr John Reinhold Forster and his son Mr George Forster were to embark with me, gentlemen skilled in natural history and botany' gave him time to 'take leave of my Family.' They sailed for Plymouth and there Lord Sandwich and Palliser went aboard for a final inspection. 'It is owing to the perseverance of these two persons that the expedition is in so much forwardness, had they given way to the general clamour and not steadily adhered to their own better judgement, the Voyage in all probability would have been laid aside. After a stay of something more than an hour they took their leave and we gave his lordship three cheers at parting.'

'Everything being at length completed, we on Monday the 13th July at 6 o'clock in the morning left Plymouth Sound with the *Adventure* in company and stood to the SW with the wind at NW, where I shall leave them and for the information of the curious give some account in what manner they are equipped.'

An account of the complement and stores is given in some detail and he concludes: 'From this general view of the equipment, the impartial reader who is a judge of marine affairs will probably conclude with me that whatever may be the event of the expedition, the ships are both well chosen and well provided.'

This second voyage had several aims, chief of which was to search for the mysterious southern continent in the manner Cook had suggested. It was also intended to test the newly designed timepieces constructed under Harrison's design for the accurate determination of longitude at sea. They took with them 'Mr Kendall's watch and three of Mr Arnold's.' Two of Arnold's were carried on the *Adventure* and the other two remained on the *Resolution.* 'The Commander, First Lieutenant and Astronomer on board each of the sloops had each of them keys of the boxes which contained the watches and were always to be present at the winding them up and comparing the one with the other.'

As soon as they got well out to sea Cook mustered the sloop's company and it was discovered that 'we have one man

SHIPS NEEDED MEN – and plenty of them. This picture of H.M.S. *Intrepid* being reviewed amply illustrates the enormous number of people required to man a naval sailing ship.
(Picture kindly supplied by Radio Times Hulton Picture Library.)

more than our complement.' Perhaps he was able to fill the place occupied by one or other of those ghostly names which again occur on the muster roll, James and Nathaniel Cook.

At Madeira 'the sloops were supplied with fresh beef and onions, and a thousand bunches of the latter were distributed among the people for a sea store. A custom I observed last voyage and had reason to think they received great benefit therefrom.'

They called in at Porto Praya in the Cape de Verde Islands to take in water and more stores, but 'the goats are so extra-ordinarily lean that hardly anything can equal them. Indeed, none of the other animals are overburdened with fat.' The governor was 'Portugese and I think has the rank of captain. He, however, lives in no sort of state, nay, everything about him has the appearance of poverty, he has a few black soldiers who are kept here as a guard, whose ludicrous appearance an Englishman cannot help laughing at, although by such behaviour

you offend them very much.' They invited the governor to
dinner on board but he, perhaps ashamed at the non-appearance
of promised bullocks, failed to turn up and tardily sent a
message excusing himself, so they 'therefore sat down to what
was on the table, piqued more at being kept so long from our
dinners than the disappointment of his company.'

Punishments on this second voyage were rare, but on one
occasion 'some petty thefts having lately been committed in
the ship, I made a thorough search today for the stolen things
and punished those in whose custody they were found.' This
disciplinary measure was followed by another, for Cook
examined the people's hands and stopped the grog of those who
had dirty ones. Most of the men were volunteers, many had
already sailed in the Endeavour and one or two had been with
Cook even before that. A stowaway came on the *Adventure*
at Madeira and was officially taken on to the ship's company
after one of their number had unfortunately died. A migrant
bird was another stowaway. 'A swallow disappeared today
which has accompanied us for several days past; it was so tame
that it came in and out of our cabins.'

On 9th September 1772 'after it was known that we were
south of the Line or Equator, the ancient custom of ducking
&c was observed, and in the evening the people were made not
a little merry with the liquor given them by the gentlemen on
this occasion.'

Six weeks later they 'anchored in Table Bay (the *Adventure*
in company). . . . At this time we have not one man on the
sick list. The people in general have enjoyed a good state of
health ever since we left England. Last night while we were off
Penguin Island the whole sea became all at once illuminated, or
what the seamen call all on fire. This appearance of the sea in
some degree is very common, but the cause is not so generally
known. Mr Banks and Dr Solander had satisfied me that it was
occasioned by sea insects. Mr Forster, however, seemed not to
favour this opinion, I therefore had some buckets of water
drawn up from alongside the ship, which we found full of an
innumerable quantity of small globular insects, about the size
of a common pin's head and quite transparent. Mr Forster, whose
province it is more minutely to describe things of this nature,
was now well satisfied with the cause of the sea's illumination.'

From the Cape Cook sent letters home, one of which was to
John Walker in Whitby:

Having nothing new to communicate, I should hardly have troubled you with a letter, were it not customary for men to take leave of their friends before they go out of the world; for I can hardly think myself in it, so long as I am deprived from having any connections with the civilised part of it and this will soon be my case for two years at least. When I think of the inhospitable parts I am going to, I think the voyage dangerous, I however enter upon it with great cheerfulness. Providence has been very kind to me on many occasions, and I trust in a continuation of the divine protection. I have two good ships, well provided and well manned. You must have heard of the clamour raised against the Resolution before I left England, I can assure you I never set foot in a finer ship. Please to make my best respects to all Friends at Whitby and believe me to be, with great regard and esteem, your most affectionate friend,

Jams. Cook

The support ship, the *Adventure,* was commanded by Tobias Furneaux who, like Drake, was a Devon man. He was seven years younger than Cook and had the same length of service, having joined the navy in 1755. He had served in various ships and was Wallis's second lieutenant in the *Dolphin.* He is described by George Robertson as 'a gentle, agreeable, well behaved good man, and very humane to all the ship's company.' His first lieutenant was Joseph Shank, who suffered from gout and at the Cape he 'having been in an ill state of health ever since we left England and not recovering here, requested my leave to quit in order to return home for the re-establishment of his health. His request appearing to be well founded, I gave him leave accordingly.' In his letter to the Admiralty on this subject Cook says: 'Mr Shank has quitted the sloop with the greatest reluctance and nothing but his bad state of health would have obliged him to give up a Voyage on which he had set his heart.'

The naturalist, Forster, 'met with a Swedish gentleman here, one Mr Sparman, who understood something of Botany and Natural History. and who was willing to embark with us. Mr Forster, thinking that he would be of great assistance to him in the course of the voyage, strongly importuned me to take him on board, which I accordingly did.'

Leaving the Cape the course was set southwards, 'the course I now intend to steer until I meet with interruption.' Soon they found 'the air begins to be pinching cold.' Fearnought jackets were issued and Cook 'set all the tailors to work to make caps to shelter them from the severity of the weather, having ordered a quantity of red baize to be converted to that purpose.' These, 'together with an additional glass of brandy every morning, enables them to bear the cold without flinching.' Strict water regulations were enforced, for they knew not how long it might be until they met with land again. A sentry was put on duty by the water casks and the Captain set an example by washing and shaving in salt water.

The thermometer was 'generally at the freezing point and sometimes below it. Rigging and sails hung with icicles. Many whales playing about the ship.' The intense cold made 'great destruction among our hogs, sheep and poultry, not a night passes without some dying. With us, however, they are not wholly lost, for we eat them notwithstanding.'

Soon they ran into ice and we learn that Cook had 'two men on board that have been in the Greenland trade. The one of them was in a ship that lay nine weeks and the other in one that lay six weeks in this kind of ice, which they call packed ice.' They 'hoisted out three boats and took up as much ice as yielded about 15 tons of fresh water. ... The melting of the ice is a little tedious and takes up some time, otherwise this is the most expeditious way of watering I ever met with.' This gave them 'an opportunity to wash and dry their linen &c, a thing that was not a little wanting.'

After this Cook 'did not hesitate one moment whether or no we should steer farther to the south, but directed my course South East by South, and as we had once broke the ice I did not doubt of getting a supply of water whenever I stood in need.' His scientific curiosity was aroused: 'Some curious and interesting experiments are wanting to know what effects cold has on sea water in some of the following instances: Does it freeze or does it not? If it does, (of which I make no doubt) what degree of cold is necessary and what becomes of the salt brine? for all the ice we meet with yields water perfectly sweet and fresh.'

Although they had two extremely reliable chronometers aboard, Cook by no means relied entirely upon them. On this second voyage he makes a further note of the excellence of

Maskelyne's method:—

> We certainly can observe with greater accuracy
> when the ship is sufficiently steady which, however,
> very seldom happens, so that most observations at
> sea are made without, but let them be made either
> the one way or the other, we are sure of finding a
> ship's place at sea to a degree and a half, and generally
> to less than half a degree. Such are the improvements
> navigation has received from the astronomers of this
> age, by the valuable table they have communicated to
> the public under the direction of the Board of
> Longitude contained in the Astronomical Ephemeris
> and the tables for correcting the apparent distance of
> the moon and a star from the effects of refraction and
> parallax; by these tables the calculations are rendered
> short beyond conception and easy to the meanest
> capacity and can never be enough recommended to
> the attention of all sea officers, who now have no
> excuse left for not making themselves acquainted
> with this useful and necessary part of their duty.

Christmas Day, 1772, was spent in the latitude of 58° South.
'At noon, seeing that the people were inclinable to celebrate
Christmas Day in their own way, I brought the sloops under a
very snug sail lest I should be surprised with a gale wind with
a drunken crew. This action was, however, unnecessary, for the
wind continued to blow in a gentle gale and the weather such
that, had it not been for the length of the day, one might have
supposed themselves keeping Christmas in the latitude of 58°
North for the air was exceedingly sharp and cold.'
    In the middle of January, 1773 they 'crossed the Antarctic
Circle, for at noon we were by observation four and a half miles
south of it and are undoubtedly the first and only ship that
ever crossed that line.' Their only companions were the grey
albatross. 'Some of the seamen call them Quaker birds, from
their grave colour.' There were icebergs, but Cook calls them
'ice islands'.*

---

*(The name of iceberg was at that time given to glacier-like formations and it was
not until 1820 that they were thus called. See *Scoresby — An Account of the Arctic
Regions* p.250)

As they reached their southern extremity 'the ice was so thick
and close that we could proceed no further but were fain to
tack and stand from it.' Not content with the report of the man
at the mast head, Cook went up himself. 'From the mast head
I could see nothing to the southward but Ice.'

They changed the course to north and then ran into 'dirty,
hazy weather with snow . . . the wind was at South a gentle
gale, but the high NE sea still kept up, which is no sign of the
vicinity of land in that quarter, yet it is there we are to expect
to find it.' Icebergs abounded, and Cook did a rough calculation
estimating the amount of ice which probably covered the sea.
'The whole quantity of ice collected together would occupy a
space of about 23,002 square miles.'

For a month they beat about, searching for the promised
land and 'various were the opinions among the officers of its
situation. Some said we should find it to the east, others to the
north, but it was remarkable that not one gave it as his opinion
that any was to be found in the south, which served to convince
me that they had no inclination to proceed any farther that
way. I, however, was resolved to get as far to the south as I
conveniently could without losing too much easting, although I
must confess I had little hope of meeting with land.' But before
very long 'the ice islands were now so numerous that we had
passed upwards of sixty or seventy since noon, many of them a
mile or a mile and a half in circuit, increasing both in number
and magnitude as we advanced to the south. . . . These obstacles,
equally as dangerous as so many rocks, together with dark nights
and the advanced season of the year, discouraged me from
carrying into execution a resolution I had taken of crossing the
Antarctic circle once more, accordingly at 4 am we tacked and
stood to the north.'

'We have a breeding sow on board which yesterday morning
farrowed nine pigs, every one of which were killed by the cold
before 4 o'clock in the afternoon, notwithstanding all the care
we could take of them. From the same cause several people on
board have their feet and hands chilblained. From these
circumstances a judgement may be formed of the summer
weather we enjoy here.'

By the middle of March, 1773, 'the moderate, and I might
almost say pleasant weather we have had at times for these two
or three days past made me wish I had been a few degrees of
latitude farther south, and even tempted me to incline a little

with our course that way, but we soon had weather which
convinced me that we were full far enough and that the time
was approaching when these seas were not to be navigated
without enduring intense cold which, however, by the by we
were pretty well used to'. He resolved to make 'the best of my
way to New Zealand.'

'If the reader of this journal desires to know my reasons for
taking the resolution just mentioned I desire he will only
consider that after cruising four months in these high latitudes it
must be natural for me to wish to enjoy some short repose in a
harbour where I can procure some refreshments for my people,
of which they begin to stand in need of, to this point too great
attention could not be paid as the voyage is but in its infancy.'

CHAPTER 8

THE
RESOLUTION ON ITS OWN

'I had the satisfaction to find that not a man was dejected, or thought the dangers we had yet to go through were in the least increased by being alone, but as cheerfully proceeded . . . wherever I thought proper to lead them, as if she . . . had been in our company.'

By the end of March 1773 they reached Dusky Bay, New Zealand, although 'in this Bay we were all strangers, in my last voyage I did no more than discover it.' They had been '117 days at sea, in which time we have sailed 3660 leagues without once having sight of land.'

'After such a long continuance at sea in a high southern latitude it is but reasonable to think that many of my people would be ill of the scurvy. The contrary, however, happened. We had only one man on board that could be called ill of this disease, occasioned chiefly by a bad habit of body and a complication of other disorders. We, are not to attribute the general good state of health in the crew wholly to the sweet wort and marmalade, this last was only given to one man; we must allow portable broth and Sour Krout to have had some share in it. This last article can never be enough recommended.'

Spruce beer, the drink which Lind had noted proved so effective in barren Newfoundland, was also brewed. Captain Cook 'used the leaves and branches of a tree which resembles the American black spruce' to make it. 'It may not be amiss to inform the reader that I have made several trials of it since we left the Cape of Good Hope, and find it to answer in a cold climate beyond all expectation.'

In Dusky Bay the crew were able to relax and caught fish and other game. 'Today we had an excellent dinner on fish, seal and wild fowl.' Cook 'employed myself in surveying, where I also met with good sport among the curlews or black birds.' They saw some natives watching them in a hostile fashion, but 'at length I landed, went up and embraced one and presented him with such articles as I had about me, which dissipated his fears and presently after we were joined by the two women. The gentlemen were with me and some of the seamen and we spent about half an hour in chit chat which was little understood on either side, in which the younger of the two women bore by far the greatest share.' They later met more of the natives and a chief who presented Cook 'with a piece of cloth and some other trifles and immediately after expressed a desire for one of our Boat Cloaks. I took the hint and ordered one to be made for him of red baize as soon as I came on board.'

When the natives came off to the ship in a canoe they were too timid to come aboard and Cook 'caused the bagpipes and fife to be played and the drum to be beat, this last they admired most, nothing, however, would induce them to come aboard.'

A few days later 'a shooting party made up of the officers went over to the north side of the bay, having the small cutter to convey them from place to place.' They managed to get themselves marooned on a small island and lost their boat. Cook came along in the nick of time and laid hold of the drifting boat 'the very moment she was going to be dashed against the rocks.' Since the tide was out they could not rescue the party until it came in at three in the morning, so they landed 'upon a naked beach, not knowing where to find a better place, and after some time got a fire and broiled some fish on which we made a hearty supper without any other sauce than a good appetite. After this we laid down to sleep having a stony beach of a bed and the canopy of heaven for a covering.'

Cook chose an uninhabited, secluded cove in which to leave some geese where 'I make no doubt but that they will breed and may in time spread over the whole country, which will answer the intent of the founder.' But 'last night the ewe and ram I had with so much care and trouble brought to this place, died, we did suppose that they were poisoned by eating of some poisonous plant, thus all my fine hopes of stocking this country with a breed of sheep were blasted in a moment.' He also 'planted several sorts of seeds.' He gives a detailed description of Dusky Bay for it may 'not only be acceptable to the curious reader but may be of use to some future navigators, for we can by no means tell what use future ages may make of the discoveries made in the present.' He ends the account by giving a description of the tea plant which they found and concludes: 'It is the business of voyagers to pass over nothing that may be useful to posterity, and it cannot be denied that these would be if ever this country is settled by a civilised people or frequented by shipping.'

From Dusky Bay the *Resolution* made for the pre-arranged rendezvous at Queen Charlotte's Sound and there found the *Adventure* awaiting them. 'That I might not idle away the whole winter in port I proposed to Captain Furneaux to spend that time in exploring the unknown parts of the sea to the east and north.'

'Knowing that celery and scurvy grass and other vegetables were to be found in this Sound and that when boiled with wheat or peas and portable soup make a very nourishing and wholesome diet which is extremely beneficial both in curing and preventing scurvy, I went myself at daylight in the morn in search of some, and returned by breakfast with a boat load, and having satisfied myself that enough was to be got, I gave orders that it should be boiled with wheat or oatmeal and portable soup for the crew of both sloops every morning for breakfast, and also with peas every day for dinner and I took care that this order was punctually complied with, at least in my sloop.' Later events were to prove the immense importance of Cook's personal supervision of this order. Two months later, when the two ships met once more and Cook 'sent aboard the *Adventure* to enquire into the state of her crew, I learnt that her cook was dead and about twenty more were attacked with the scurvy.' We find it significant that it was the cook himself who had died, for he was the very man who was chiefly responsible for serving the recommended greens and it 'was a new diet, which alone was sufficient for seamen to reject it. To introduce any new article of food among seamen, let it be ever so much for their own good, requires both the example and authority of a commander, without both of which it will be dropped before the people are sensible of the benefits resulting from it; was it necessary I could name fifty instances in support of this remark. Many of my people, officers as well as seamen, at first disliked celery, scurvy grass &c being boiled in the peas and wheat, and some refused to eat it, but as this had no effect on my conduct, this obstinate kind of prejudice, by little and little, wore off and they began to like it as well as the others, and now, I believe, there was hardly a man in the ship that did not attribute our being so free of the scurvy to the beer and vegetables we made use of in New Zealand.'

In these few journal entries lies the whole crux of the success of Cook's voyages. Without a healthy crew he could have done very little. It was this seemingly small action of his own example and insistence which ensured the success of the whole venture. 'After this I seldom found it necessary to order any of my people to gather vegetables whenever we came where any was to be got, and if scarce, happy was he who could lay hold of them first.'

The natives in Queen Charlotte's Sound brought them some fish and 'we had their company at breakfast. One of them agreed to go away with me, but he afterwards was of another mind . . . It was not uncommon for them to bring their children with them aboard and present them to us in expectation of our making them presents. This happened to me yesterday morning, a man brought his son, a boy about 10 years of age and presented him to me.' At first Cook thought he wanted to sell his son but then he 'found out that he wanted me to give him a shirt, which I accordingly did. The boy was so fond of his new dress that he went all over the ship presenting himself to everybody that came his way. The liberty of the boy offended old Will, the ram goat, who up with his head and knocked the boy backwards on the deck. Will would have repeated his blow had not some of the people got to the boy's assistance, this misfortune however, seemed to him to be irreparable, the shirt was dirtied and he was afraid to appear in the cabin before his father until brought in by Mr Forster, when he told a very lamentable story against Goure the great Dog, for so they call all the quadrupeds we have aboard, nor could he be pacified till his shirt was washed and dried.'

The following extract from the *Journal* of this time is of special interest in showing Cook's capacity for objective judgement and his acute and far seeing appraisal of the consequences of European intrusion into the social life and economy of primitive peoples. A point of view is expressed here which time only has made obvious, and it serves to illustrate most forcibly the stature of Cook and the quality of his mind.

'During our short stay in this Sound I have observed that this second visit of ours hath not mended the morals of the natives of either sex. The women of this country I always looked upon to be more chaste than the generality of Indian women. Whatever favours a few of them might have granted to the crew of the *Endeavour*, it was generally done in a private manner and without the men seeming to interest themselves in it, but now we find the men are the chief promoters of this vice, and for a spike nail or any other thing they value will oblige their wives and daughters to prostitute themselves whether they will or no, and that with not the privacy decency seems to require. Such are the consequences of a commerce with Europeans, and what is still more to our shame, civilised Christians. We debauch their morals, already too prone to vice, and we introduce among them wants and perhaps diseases which they never before knew and which serve only to destroy that happy tranquility they and their forefathers had enjoyed. If any one denies the truth of this assertion, let him tell me what the natives of the whole extent of America have gained by the commerce they have had with Europeans.'

At the beginning on June 1773 'the wind coming more favourable, we unmoored and at 7 weighed and put to sea with the *Adventure* in company.' There were no untoward happenings on this part of the voyage, unless one can count: 'One of our goats fell overboard, hoisted a boat out and took it up alive, but it died soon after.'

On the 17th July 1773 Cook observed that they had 'run down the whole of the longitude I at first intended' and so he 'steered N½E having the advantage of a strong gale at SSW, with a view of exploring that part of the sea down as low as the latitude of 27° in which space no one has been that I know of.' At the beginning of August he writes: 'It will hardly be denied but what I must have formed some judgement concerning the great object of my researches, viz, the Southern Continent. Circumstances seem to point out to us that there is none, but this is too important a point to be left to conjecture; facts must determine it and these can only be had by visiting the remaining unexplored parts of this sea, which will be the work of the remaining part of this voyage.'

Once again they made for Tahiti to refresh themselves. There, the Resolution was very nearly wrecked on a reef. Cook 'had given directions in what position the land was to be kept, but by some mistake it was not properly attended to, for when I got up at break of day I found we were steering a wrong course and were not more than half a league from the reef which guards the south end of the island.' The *Adventure*, however, had kept clear and she 'got under sail with the land wind.' Eventually 'the sloops were got once more into safety after a narrow escape of being wrecked on the very island we but a few days ago so ardently wished to be at.'

The first thing they did on landing was to get some water, 'having scarce any left on board.' When they anchored in Matavai Bay 'many of the natives came off to us, several of whom I knew and almost all of them me.' The following morning 'I had no sooner landed than I was met by a venerable old lady, mother of the late Toutaha, she seized me by both hands and burst into a flood of tears saying 'Toutaha Tiyo no Toute matte' (Toutaha the friend of Cook is dead). I was so much affected at her behaviour that it would not have been possible for me to refrain mingling my tears with hers, had not Otoo come and snatched me, as it were, from her.'

Cook entertained the King 'with the bagpipes, of which music he was very fond, and dancing by the seamen. He in return ordered some of his people to dance also, which dancing consisted chiefly in strange contortions of the body, there were some of them that could, however, imitate the seamen tolerably well, both in country dances and hornpipes.'

They stayed only a fortnight in Tahiti and on the 1st September, 1773, 'The sick being all pretty well recovered, our water casks repaired and filled and the necessary repairs of the sloops completed, I determined to put to sea without loss of time.' Although the island had previously 'swarmed as it were with hogs and fowls' they were now very scarce, so they were glad to reach Huahine where there was 'a plentiful supply of fresh pork and fowls, which to people who had been living ten months on salt meat was no unwelcome thing.'

The old chief, Oree, remembering Cook's first visit, demonstrated that 'friendship is sacred with these people' for he 'came and fell upon my neck and embraced me, this was by no means ceremonious, the tears which trickled plentifully down his cheeks sufficiently spoke the feelings of his heart.' It was here, Huahine, that Omai came aboard the *Adventure* and started his own adventure. The Resolution took a native called Odiddy; he, however, only sailed with them for about six months, returning to his native land the next time they called there. They 'procured not less than 300 hogs to both sloops besides fowls and fruit and had we stayed longer might have got many more for neither hogs nor fowls were apparently diminished, but everywhere appeared as numerous as ever, such is the state of the little but fertile Isle of Huahine.'

They called at Raiatea, where they found the same chief, Oreo who 'received us with great cordiality, expressed much satisfaction at seeing me again, desired that he might be called Cook (or Toote) and I Oreo, which was accordingly done.' Oreo 'introduced into the ship two very pretty young women. These two beauties attracted the notice of most of the officers and gentlemen, who made love to them in their turns. The ladies very obligingly received their addresses, to one they gave a kind look, to another a smile, thus they distributed their favours to all, received presents from all and at last jilted them all.' The crews of the ships' boats fared better, for they 'made the circuit of the isle and were hospitably entertained by the people who provided them with victuals, lodging and bedfellows according to the custom of the country.'

But, 'some injustice has been done the women of Tahiti and the Society Isles by those who have represented them without exception as ready to grant the last favour to any man that will come up to their price. But this is by no means the case; the favours of married women, and also the unmarried of the better sort, are as difficult to obtain here as in any other country whatever. . . . That there are prostitutes here as well as in other countries is very true, perhaps more in proportion, and such were those that came on board the ship to our people and frequented the post we had on shore. . . . The truth is that the woman who becomes a prostitute does not seem on their opinion to have committed a crime of so deep a die as to exclude her from the esteem and society of the community in general. On the whole, a stranger who visits England might with equal justice draw the characters of the women there from those he might meet with on board the ships in one of the naval ports, or in the purlieus of Covent Garden & Drury Lane.'

When the *Resolution* sailed the Chief 'took a very affectionate leave and asked me if and when I would return; these were his last words, questions which have very often been put to me by many of these islanders, and whenever this happened they never failed to tell me to bring my sons with me, for they would frequently ask me how many children I had and whether they were boys or girls &c. My friend Oree of Huahine was very desirous for me to return to his isle, but as he did not expect that this could be done till the expiration of the same time as I had been absent before, he one time very justly observed that both he and I might be dead but, says he: 'Let your sons come, they will be well received.' In regard to the Polynesian languages, Cook was an assiduous student. All his journals contain vocabularies of words which he had learnt from the natives, and he was also adept at communication by signs, a method in which he became very 'fluent'.

A week later Cook passed and named the Sandwich Isles, 'in honour of my noble patron, the Earl of Sandwich.' By October 2nd, 1773 they had reached Tonga, where they were 'welcomed ashore by acclamations from an immense crowd of men and women, not one of which had so much as a stick in their hands.' The chief invited them to his house where Cook 'ordered the bagpipes to be played and in return the chief ordered three young women to sing a song, which they did with a very good grace. When they had done, I gave each of them a necklace, this set most of the women in the circle a singing; their songs were musical and harmonious, noways harsh or disagreeable.' Bananas and cocoanuts were set before them and 'a bowl of liquor prepared in our presence of the juice of Eava for us to drink. Pieces of the root were first offered us to chew, but as we excused ourselves from assisting in the operation, it was given to others to chew, which done, it was put into a large wooden bowl and mixed with water in the manner already related, and as soon as it was properly strained for drinking, they made cups of green leaves which held near half a pint and presented to each of us a cup of the liquor, but I was the only one who tasted of it, the manner of brewing had quenched the thirst of everyone else.'

The bartering began to get out of hand, for the natives brought out only 'cloth and other curiosities, things which I did not come here for and for which the seamen only bartered away their clothes. In order to put a stop to this and to obtain the refreshments we wanted, I gave orders that no curiosities should be purchased by any person whatever. . . . This had the desired effect for in the morning the natives came off with bananas and cocoanuts in abundance and some fowls and pigs which they exchanged for nails.' After they left Cook estimated that all in all 'they have not got less than three or four hundred weight from the largest spike down to a sixpenny nail.'

They visited a chief's house inland where they were 'no sooner seated than the oldest of the priests began a speech or prayer which was first addressed to the Afia-tou-ca (shrine) and then to me and it alternatively. When he addressed me he paused at each sentence till I gave a nod of approbation. I, however, did not understand one single word of what he said. At times the old man seemed to be at a loss what to say, or perhaps his memory failed him, for every now and again he was prompted by another who sat by him.' On a second visit to the same place they 'had no praying, but on the contrary here the good natured old chief introduced me to a woman and gave me to understand that I might retire with her; she was next offered to Captain Furneaux but met with a refusal from both, though she was neither old nor ugly.' They were constantly offered Eava to drink and with great perception Cook comments on this: 'One would not wish for a better sign of friendship than this: can we make a friend more welcome than by setting before him the best liquor in our possession or that can be got?'

He noticed their courtesy: 'Everything you give them they apply to their heads by way of thanks. This custom they are taught from their infancy; when I have given things to little children the mother has lifted up the child's hands to its head just as we in England teach a child to pay a compliment by kissing its hand. ... The women have very often, when I have been among them, taken hold of my hand, kissed it and then laid it to their heads.' The women of Tonga he found to be 'the merriest creatures I ever met with, and will keep chattering by one's side without the least invitation or consideration whether or no they are understood, provided one does but seem pleased with them. In general they appeared to be modest, though there were no want of those of a different stamp.'

As they left Tonga, 'a canoe conducted by four men came alongside with one of those drums already mentioned on which one man kept continually beating, thinking no doubt that we should be charmed with his music. I gave them a piece of cloth and a nail for their drum.'

The two ships reached New Zealand at the end of October, but 'the sea rose in proportion with the wind, so that we not only had a furious sea but a mountainous sea also to encounter, thus after beating two days against strong gales and arriving in sight of our port we had the mortification to be driven off from the land by a furious storm. A week later they were in hopes of 'reaching Queen Charlotte's Sound on the next flood tide. Vain were our expectations ... we lost more on the ebb than we had gained on the flood.'

On the 4th November 1773 they managed to land and 'set up

tents for the reception of the sail makers, coopers and others whose business made it necessary for them to be on shore; began to caulk the decks and sides, overhaul the rigging, to cut firewood and set up the forge to repair the iron work, all of which were absolutely necessary occupations.' Several of the natives 'made us a visit, among whom were some that I knew when I was here in the *Endeavour*.'

Cook learnt that the pigs which were left by Captain Furneaux had been taken away and 'the two goats have likewise been caught, killed and eaten; thus all our endeavours for stocking this country with useful animals are likely to be frustrated by the very people whom we meant to serve. Our gardens had fared somewhat better, everything in them except the potatoes, they had left entirely to nature, who had acted her part so well that we found most articles in a flourishing state. The potatoes they had dug up, some few, however, remained and were growing.'

On his first voyage Cook had stated that there were cannibals in New Zealand. He now had direct evidence to prove this for 'some of the officers went on shore to amuse themselves among the natives when they saw the head and bowels of a youth who had lately been killed; the heart was stuck upon a forked stick and fixed to the head of their largest canoe. The gentlemen brought the head on board with them. I was on shore at this time but soon after returned on board where I was informed of the above circumstance and found the quarter deck crowded with natives. I now saw the mangled head . . . a piece of the flesh had been broiled and eaten by one of the natives in the presence of most of the officers. The sight of the head and the relation of the circumstances just mentioned struck me with horror and filled my mind with indignation against these cannibals, but when I considered that any resentment I could show would avail but little, and being desirous of being an eye witness to a fact which many people had their doubts about, I concealed my indignation and ordered a piece of the flesh to be broiled and brought on the quarter deck, where one of these cannibals ate it with a seeming good relish before the whole ship's company, which had such an effect on some of them as to cause them to vomit. . . . That the New Zealanders are cannibals can no longer be doubted.'

As well as the vegetables which had grown from their seeds they 'found everywhere plenty of scurvy grass and celery, which

I caused to be dressed every day for all hands, by this means they have been mostly on a fresh diet for these three months past and at this time we had neither a sick or scorbutic person on board.'

There was no sign of the *Adventure*. Cook was, 'however, under no sort of anxiety for her safety. . . . It is not likely that we shall join again, as no rendevous was absolutely fixed upon after leaving New Zealand. Nevertheless, this shall not discourage me from fully exploring the southern parts of the Pacific Ocean.'

'On our quitting the coast and consequently all hopes of being joined by our consort, I had the satisfaction to find that not a man was dejected, or thought the dangers we had yet to go through were in the least increased by being alone, but as cheerfully proceeded to the south or where ever I thought proper to lead them as if she or even more ships had been in our company.' The course was set southwards and the little ship sailed into those uncharted waters inhabited by 'seals, penguins, albatrosses, petrels and whales.'

By December 7th, 1773, 'we reckoned ourselves Antipodes to our friends in London, consequently as far removed from them as possible.' The following week they saw their first 'island of ice' and after that 'the ice begins to increase fast. From noon till 8 o'clcok in the evening we saw but two islands, but from 8 to 4 we passed fifteen.' One iceberg 'was near proving fatal to us, we had not weathered it more than once or twice our length, had we not succeeded, this circumstance could never have been related. According to the old proverb a miss is as good as a mile, but our situation requires more misses than we can expect. This, together with the improbability of meeting with land to the south and the impossibility of exploring it for the ice if we did find any, determined me to haul for the north.' The weather was 'dark and gloomy and very cold, our sails and rigging hung with icicles for these two days past.'

Nevertheless, within a week, 'the wind veering to the NW and the clear weather tempted me to stand to the south, which we accordingly did.' Once again they ran into 'such a vast quantity of field or loose ice as covered the whole sea from south to east and was so thick and close as to obstruct our passage.' They took in a good deal of it which was melted for drinking water. While they were taking up the ice 'two of the Antarctic petrels so often mentioned were shot . . . They were both casting their feathers, and yet they were fuller of them than any birds we had seen, so much has nature taken care to clothe them suitable to the climate in which they live.' The weather was 'piercing cold' and once again they steered north. Christmas Day passed with no special note in the *Journal*, but the naturalist Forster wrote in his journal: 'This being Christmas Day the Captain, according to custom, invited the officers and mates to dinner and the lieutenants entertained the petty officers. The sailors feasted on a double portion of pudding, regaling themselves with the brandy of their allowance, which they had saved for the occasion some months beforehand, being solicitous to get very drunk, though they are commonly solicitous about nothing else.'

The chronometer was proving its worth. 'At 9 o'clock had again several observations of the sun and moon. The results were confirmable to yesterday and determined our longitude beyond a doubt. Indeed, our error can never be great so long as we have so good a guide as Mr Kendall's watch.'

A month later they 'came the third time within the Antarctic Polar Circle.' They saw several whales and 'two ice islands, one of which was very high, terminating in a peak like the cupola of St Paul's Church, we judged it to be 200 feet high.' The course was southwards once again. 'Presently came on a thick fog, this made it unsafe to stand on, especially as we had seen more ice islands ahead; we therefore tacked and made a trip to the north for about one hour and a half, in which time the fog dissipated and we resumed our course to the SSE.' They came to a vast ice field where they 'counted ninety-seven ice hills or mountains, many of them vastly large. Such ice mountains as these are never seen in Greenland, so that we cannot draw a comparison between the Greenland ice and this now before us.'

It is here that Cook's momentous and oft-repeated words occur: 'I will not say that it was impossible anywhere to get farther to the south, but the attempting of it would have been a dangerous and rash enterprise, and what I believe no man in my situation would have thought of. It was indeed my opinion, as well as the opinion of most on board, that this ice extended quite to the Pole, or perhaps joins to some land to which it had been fixed from the creation. . . . I, who had ambition not only to go farther than anyone had done before, but as far as it was possible for man to go, was not sorry at meeting with this interruption, as it in some measure relieved us, at least shortened the dangers and hardships inseparable with the navigation of the southern Polar regions. Since therefore we could not proceed

one inch farther, no other reason need be assigned for my tacking
and standing back to the north.'

CHAPTER 9

THE PACIFIC

'They were so far from wishing the voyage at an end
that they rejoiced at the prospect of its being
prolonged another year.'

By the beginning of the year 1774, Cook thought that all the
conjectures of a southern continent could now be 'entirely
refuted, as all their enquiries were confined to this Southern
Pacific Ocean in which, although there lies no continent there
is, however, room for very large islands, and many of those
formerly discovered are very imperfectly explored and their
situations as imperfectly known.' 'All these things considered,
and more especially as I had a good ship, a healthy crew and no
want of stores or provisions, I thought I could not do better
than spend the ensuing winter within the tropics.'

We have here a good example of Captain Cook writing his
intentions down and we are able to see how well he adhered to
his original plan. 'My intention was first to go in search of the
land said to have been discovered by Juan Fernandez ... If I
failed of finding this land, then go in search of Easter Island,
whose situation is known with so little certainty that the
attempts lately made to find it have miscarried. I next intended
to get within the Tropic and then proceed to the west, touching
at and settling the situations of such isles as we might meet
with, till we arrived at Tahiti, where it was necessary I should
touch to look for the *Adventure*. I had also thought of running
as far west as Tierra Austral del Espiritu Santo, discovered by
Quiros and which Bougainville calls the Great Cyclades, ...
intending if possible to be the length of Cape Horn in November
next, when we should have the best part of the summer before
us to explore the southern part of the Atlantic Ocean.'

'Great as this design appeared to be, I however thought it was
possible to be done, and when I came to communicate it to the
officers I had the satisfaction to find that they all heartily
concurred in it. I should not do my officers justice if I did not
take some opportunity to declare that they always showed the

utmost readiness to carry into execution in the most effectual manner every measure I thought proper to take. Under such circumstances it is hardly necessary to say that the seamen were always obedient and alert, and on this occasion they were so far from wishing the voyage at an end that they rejoiced at the prospect of its being prolonged another year and soon enjoying the benefits of a milder climate.'

Not finding the alleged island of Juan Fernandez, they gave up the search and 'stood away to the north in order to get into the latitude of Easter Island.'

It was at this time that Cook was 'taken ill of the bilious colic and so violent as to confine me to my bed, so that the management of the ship was left to Mr Cooper, my First Officer, who conducted her very much to my satisfaction. It was several days before the most dangerous symptoms of my disorder were removed, during which time Mr Patten the Surgeon was to me not only a skilled physician, but a tender nurse and I should ill deserve the care he bestowed on me if I did not make this public acknowledgement. When I began to recover, a favourite dog belonging to Mr Forster fell a sacrifice to my tender stomach. We had no other fresh meat whatever on board, and I could eat of this flesh, as well as broth made of it, when I could taste nothing else, and thus I received nourishment and strength from food which would have made most people in Europe sick, so true it is that necessity is governed by no law.'

Easter Island was sighted on 11th March, 1774. They landed at a sandy beach where about a hundred natives were collected. 'Not one of them had so much as a stick or weapon of any sort in their hands. After distributing a few trinkets among them we made signs for something to eat, on which they brought down a few potatoes, plantains and sugar cane and exchanged for nails, looking glasses and pieces of cloth. We presently discovered that they were as expert thieves and as trickish in their exchanges as any people we had yet met with. It was with some difficulty we could keep the hats on our heads, but hardly possible to keep anything in our pockets, not even what themselves had sold us, for they would watch every opportunity to snatch it from us, so that we sometimes bought the same thing two or three times over and after all did not get it.'

Odiddy 'understood their language (though but very imperfectly) much better than any of us.' They stayed only a day or two and Cook concluded 'no nation will ever contend for the honour of the discovery of Easter Island as there is hardly an island in this sea which affords less refreshments and conveniences for shipping than it does. Nature has hardly provided it with any thing fit for man to eat or drink, and as the natives are but few and may be supposed to plant no more than sufficient for themselves, they cannot have much to spare to new comers.'

April brought them to the Marquesas, where they traded for pigs, but unfortunately the trading was soon spoilt by one of the 'young gentlemen giving for a pig a very large quantity of red feathers he had got at Amsterdam, which these people much value and which the other did not know, nor did I know at this time that red feathers was what they wanted.'

When Cook saw 'that this place was not likely to supply us with sufficient refreshments, not very convenient for getting off wood and water, nor for giving the ship the necessary repairs, I resolved forthwith to leave it and seek for some place that would supply our wants better, for it must be supposed that after having been 19 weeks at sea (for I cannot call the two or three days spent at Easter Island anything else) living all the time on a salt diet, but what we must want some refreshments. Although I must own, and that with pleasure, that on our arrival here it could hardly be said that we had one sick man on board, and not above two or three who had the least complaint. This was undoubtedly owing to the many anti-scorbutic articles we had on board and the great care and attention of the Surgeon who took special care to apply them in time.'

They paused only long enough to name Hood's Island 'after the young gentleman who first saw it.' and then made for Tahiti. One of the reasons for Cook's putting in there 'was to give Mr Wales an opportunity to know the error of the watch from the known longitude of this place and to determine afresh her rate of going. The first thing we did was to land his instruments &c and to set up tents for the reception of a guard and such others as it was necessary to have on shore. As to sick, we had none.'

The red feathers became a valuable currency. 'Having these feathers was a very fortunate circumstance to us, for as they were valuable to the natives they became so to us also, for our stock of trade was by this time greatly exhausted and if it had not been for them I should have found it difficult to have supplied the ships with the necessary refreshments.' (Although

Cook mentions 'ships' in the plural, the *Adventure* was not at
Tahiti, but on her way home.)

They found an upsurge of building going on, 'a great number
of canoes and houses both large and small. People living in
spacious houses who had not a place to shelter themselves
in eight months ago.' 'The tools which they got from the
English and other nations who have touched here no doubt
accelerated the work. ... The number of hogs was another
thing which struck our attention. They certainly had a good
many when we were here before, but not choosing to part
with any had conveyed them out of our sight.' The red feathers
altered all that. 'All red feathers are esteemed and they are such
good judges as to know very well how to distinguish one sort
from another. Many of our people attempted to deceive them
by dyeing other feathers, but I never heard that any one
succeeded. These feathers they make up in little bunches ...
they are used in all their religious ceremonies. I have very often
seen them hold one of these bunches and say a prayer, not
one word of which I could ever understand. Whoever comes to
this isle will do well to provide himself with red feathers, the
finest and smallest that are to be got. He must also have a good
stock of axes and hatchets, spike nails, files, knives, looking
glasses, beads &c. Sheets and shirts are much sought after,
especially by the ladies as many of our gentlemen found by
experience.'

'The ladies of Tahiti, after they had pretty well cleared
their lovers of shirts found a method of clothing themselves
with their own cloth. It was their custom to go on shore every
morning and return on board in the evening, generally clad in
rags, which admitted of an excuse to importune the lover for
better clothes and when he had no more of his own, he was to
clothe them with new cloth of the country, which they always
left ashore and appeared again in rags and must again be clothed,
so that the same suit might pass through twenty different hands
and be as often sold, bought and given away.'

Though the sheep they had left had died, the goats had fared
well and multiplied. The Resolution also 'furnished them with
a stock of cats, no less than twenty were given away at this isle,
besides what were left at Ulieta and Huahine.'

The fleet, 'upwards of three hundred canoes' were assembled
and Cook went ashore to pay the chief, Otoo, a formal visit. He
took hold of Cook's right hand; another took his left. One

wished to take him to review the fleet and the other wanted to
conduct him to the king, 'so that between the one party and
the other I was like to have been torn to pieces and was obliged
to desire him to desist.'

Odiddy was greatly distressed, for the islanders wished him
to stay with them. Cook 'frankly told him that if he went to
England it was highly probable that he would never return,
but if after all he chose to go I would take care of him and he
must look upon me as his father. He threw his arms about me
and wept. . . . He was very well beloved in the ship, for which
reason everyone was persuading him to go with us, telling what
great things he would see and return with immense riches,
according to his ideas of riches. But I thought proper to
undeceive him, thinking it an act of the highest injustice to
take away a person from these isles against his own free
inclination under any promise whatever, much more that of
bringing him back again. What man on board can make such a
promise as this?'

Mr Forster persuaded Odiddy to go with them to Huahine and
he eventually stayed there when the *Endeavour* left. 'He was a
youth of good parts, and like most of his countrymen of a
docile, gentle and humane disposition. . . . Indeed, he would
have been a better specimen of the nation in every respect than
the one on board the Adventure.' This was Omai, who did
return to Britain with Captain Furneaux, and proved to be
better than Cook had first thought.

Odiddy was probably the means of saving Cook's life, for
while at Huahine his party was almost ambushed and the story
of the event is worthy of the best Hollywood tradition. It seems
that three of the officers were robbed. A message was sent to
Cook from Oree, the chief, asking him to 'come ashore with 22
men to go with him to chastise the robbers. . . . Accordingly we
landed 48 men, including myself, Mr F. and officers. The chief
joined us with a few people and we set out on our march in
good order. The chief's party gathered like a snowball as we
marched through the country, some armed and some not.
Odiddy, who was with us, began to be alarmed and told us
that many of the people in our company were of the party we
were going against, and at last told us that they were only
leading us to some place where they could attack us to advantage
. . . I told the chief that I would march no farther, for we were
then crossing a deep valley bounded on each side with steep

rocks, where a few men with stones only might have cut off
our retreat, supposing their intention to be what Odiddy had
said and what he still abided by. We returned back in the same
order as we went and saw in several places people come down
from the sides of the hills with their arms in their hands, which
they laid down whenever they found they were seen by us.'

They called at Raiatea where Cook 'went on shore accompanied by Mr F. &c to make the chief the customary present.
At our first entering his house we were met by 4 or 5 old women
weeping and lamenting, as it were, most bitterly, and at the
same time cutting their heads with instruments made of sharks'
teeth, so that the blood ran plentifully down their faces and
on to their shoulders, and what was still worse we were obliged
to submit to the embraces of these old hags, and by that means
got all besmeared with blood. This ceremony (for it was merely
such) being over, these women went and washed themselves
and immediately after appeared as cheerful as any of the
company.'

They were later 'entertained with a play which ended with
the representation of a woman in labour who at last brought
forth a thumping boy near six feet high, who ran about the
stage ... As soon as they got hold of the fellow who
represented the child, they flattened his nose or pressed it up
to his face, which may be a custom among them and the reason
why they have all in general flat, or what we call pug noses.'

At length, having 'got on board a plentiful supply of all
manner of refreshments' Cook 'directed my course to the west
and took our final leave of these happy isles on which benevolent
nature with a bountiful and lavishing hand hath bestowed
every blessing man can wish.' Two weeks later they thought
they saw land, but found it was only clouds. Eventually, at the
end of June 1774, they came to the Friendly Islands where
they were 'received with great courtesy by the natives' who,
however, stole several things, an adze among them. Cook
'addressed myself to several people to have the adze returned,
especially to an elderly woman who had always a great deal to
say to me from my first landing, but on this occasion she gave
her tongue free liberty, not one word in fifty I understood. All
I could learn from her arguments was that it was mean in me
to insist on the return of so trifling an article, but when she
found I was determined, she and 3 or 4 more women went
away and soon after the adze was brought me, but I saw her no

more, which I was sorry for as I wanted to make her a present on account of the part she seemed to take in all our transactions, private as well as public, for . . . the first time I landed this woman and a man presented to me a young woman and gave me to understand she was at my service. Miss, who probably had received her instructions, I found wanted by way of handsel a shirt or a nail, neither the one nor the other I had to give her without giving her the shirt on my back, which I was not in a humour to do. I soon made them sensible of my poverty and thought by that means to have come off with flying colours, but I was mistaken, for I was made to understand that I might retire with her on credit. This not suiting me neither, the old lady began first to argue with me and when that failed she abused me. I understood very little of what she said, but her actions were expressive enough and showed that her words were to this effect, sneering in my face and saying: What sort of a man are you thus to refuse the embraces of so fine a young woman? For the girl certainly did not want beauty, which I could, however, withstand but the abuse of the old woman I could not, and therefore hastened into the boat.'

After careful observations Cook found that 'Bougainville's discoveries are laid down about 20 miles too far north, in the charts which are bound up with the English translation. As this error is nearly equal and the same way as the one at Tahiti it is probable this error runs through the whole chart of this sea.'

At one island which they came to, Malekula, they found the natives 'quite different to all we have yet seen and speak a different language. They are almost black, or rather a dark chocolate colour, slenderly made, not tall, have monkey faces and woolly hair.' He refers to them as an 'apish nation' and says they are the 'most ugly and ill-proportioned of any that I ever saw' with thick lips and flat noses.

In the group of islands which Cook eventually named the *New Hebrides* their reception, on the whole, was 'civil and good natured when not prompted by jealousy to a contrary conduct. A conduct one cannot blame them for when one considers the light in which they must look upon us; it is impossible for them to know our real design, we enter their ports without their daring to make opposition, we attempt to land in a peaceable manner, if this succeeds it is well, if nor, we land nevertheless and maintain the footing we thus got by the superiority of our fire arms. In what other light can they at first look upon us but as invaders of their country?'

It was not only the natives who needed to be controlled. Cook also had constantly to watch his own men in order to prevent them becoming trigger-happy. On one occasion the Resolution was preparing to leave and 'a good many of the natives were, as usual, assembled at the landing place and unfortunately one of them was shot by one of our sentries. I who was present and on the spot, saw not the least cause for the committing of such an outrage and was astonished beyond measure at the inhumanity of the act. The rascal who perpetrated this crime pretended that one of the natives laid his arrow across his bow and held it in the attitude of shooting, so that he apprehended himself in danger, but this was no more than what was done hourly, and I believe with no other view than to let us see that they were armed as well as us.' The marine who fired the gun was punished by being confined for two months.

There was a volcano which Cook observed 'at no great height, nor is the volcano at the highest part of it, but on the side, and contrary to the opinion of philosophers, which is that all volcanoes must be at the summits of the highest hills. . . . Here seems to be a field open for some philosophical reasoning, but as I have no talent that way I must content myself with stating facts as I found and leave the causes to men of more ability.'

CHAPTER 10
THE CIRCUMNAVIGATION COMPLETE

'I can be bold to say that no man will ever venture farther than I have done.'

By September 1774 Cook had 'finished the survey of the whole archipelago, so that I had no more business there. Besides, the season of the year made it necessary I should think of returning south.' From the New Hebrides they stood to the south and discovered another island, Balade. It seemed to be a poor island, they 'heard the crowing of cocks but saw none, or anything else to induce us to believe they had anything to spare us but good nature and courteous treatment.'

They were 'visited by some hundreds of the natives; some came off in canoes, others swam off, so that before 10 o'clock our decks were quite full. . . . They brought with them some arms, such as clubs, darts &c, which they exchanged away. Indeed, these things generally found the best market with us, such was the prevailing passion for curiosities, or what appeared new. As I have had occasion to make this remark more than once before, the reader will think the ship must be full of such articles by this time. He will be mistaken, for nothing is more common than to give away what has been collected at one island for anything new at another, even if it is less curious. This, together with what is destroyed on board after the owners are tired of looking at them, prevents any considerable increase.'

Cook gave the chief two dogs. 'It was some time before he was satisfied the dog and the bitch was intended, but as soon as he was convinced he could hardly contain himself for joy.' They were unfamiliar with pigs and Cook gave them two. He expounded 'on the merits of the two pigs, showing them the distinction of their sex, telling them how many young ones the female would have at a time. In short, I multiplied them to some hundreds in a trice. My only view was to enhance the value of the present that they might take the more care of them, and I had reason to think I in some measure succeeded.'

At times the natives crowded round to such an extent that it was necessary to draw a line 'on the ground, within which the natives were given to understand they were not to come. One of them happened to have a few cocoanuts, which one of our people wanted to purchase, but as the other was unwilling to part with them he ran off. When he saw he was followed, he sat down on the sand and made a circle round him, as he had seen our people do, and signified none were to come within it, which was accordingly observed. As this story was well attested I thought it not unworthy of a place in this Journal.' They found the women 'of this country . . . far more chaste than those of the eastern islands. I never heard that one of our people obtained the least favour from any one of them. I have been told that the ladies here would frequently divert themselves by going a little aside with our gentlemen as if they meant to grant them the last favour, and then run away laughing at them. Whether this was chastity or coquetry I shall not pretend to determine, since the consequences were the same.'

On a distant island they saw 'an elevation like a tower. Several more appeared on different parts of the coast to the west, and so numerous that they looked not unlike the masts of a fleet. . . . Various were the opinions and conjectures about them and occasioned the laying of several trifling wagers.' Although there was a 'melancholy prospect of a sea strewn with shoals' and Cook confessed to being 'almost tired of a coast I could no longer explore' he was determined not to leave it until he had found out what they were. As they got nearer 'everyone was now satisfied that they were trees, except our philosophers, who still maintained they were stone pillars.' It is not recorded what the Forsters said when they landed and proceeded to cut them down, for they were 'a kind of spruce pine.'

Cook gave the name of *New Caledonia* to this group of islands and then steered for New Zealand in order to refresh his people before starting to 'cross this vast ocean . . . so as to pass over those parts which were left unexplored last summer.'

On the 18th October 1774 they reached New Zealand and 'hauled round *Point Jackson* through a sea which looked terrible, occasioned by the tide, wind and sea, but as we knew the cause it did not alarm us.' At first the natives were frightened, but 'the moment we landed they knew us again, joy took place of fear, they hurried out of the woods, embraced us over and over again and skipped about like mad men.'

In the light of later knowledge it is easy to understand the natives' fear, for it was here that a party of ten of the *Adventure's* crew had been murdered and eaten almost a year before this arrival of the *Resolution*. Cook could not know this. The message which he had left for Captain Furneaux was gone and he knew he had been there 'by several trees having been cut down with saws and axes, which were standing when we sailed. This ship could be no other than the *Adventure*.'

After a couple of weeks they were off once more. By the end of November, 1774 Cook 'now gave up all hope of finding any more land in this ocean, and came to a resolution to steer directly to the West entrance to the Straits of Magellan, with a view of coasting the out- or south side of Tierra del Fuego round Cape Horn to Strait La Maire. As the world has but a very imperfect knowledge of this coast I thought the coastline of it would be of more advantage to both Navigation and Geography than anything I could expect to find in a higher latitude.'

'I have now done with the SOUTHERN PACIFIC OCEAN, and flatter myself that no one will think that I have left it unexplored, or that more could have been done in one voyage towards obtaining that end than has been done in this.'

December saw them at Tierra del Fuego. Cook observed 'that this is the most desolate coast I ever saw, it seems to be entirely composed of rocky mountains without the least appearance of vegetation. These mountains terminate in horrible precipices whose craggy summits spire up to a vast height, so that hardly anything in nature can appear with a more barren and savage aspect than the whole of this coast. The inland mountains were covered with snow, but those on the sea coast were not.' One of the rocks terminated in two high towers and was thus named *York Minster*.

Just before Christmas they found a sheltered harbour and 'a good anchoring place a little to the west of this harbour, before a stream of water which comes out of a lake which is continually supplied by a cascade falling into it.' Birds were plentiful and they made up two shooting parties. 'There were abundance of geese. . . . By one method and another, we got sixty-two, with which we returned in the evening, all heartily tired; but the acquisition of so many geese over balanced every other consideration, and we sat down with a good appetite to supper.' The other party brought fourteen geese, 'so that I was able to make a distribution to the whole crew, which was the more acceptable on account of the approaching festival.' Needless to say, they named the place *Goose Island.*

Some natives came to the ship in their canoes. 'They are a little, ugly, half starved, beardless race; I saw not a tall person among them. They were almost naked, their clothing was a seal skin. . . . I saw two young children at the breast as naked as they were born, thus they are inured from their infancy to cold and hardships. They had with them bows and arrows and harpoons made of bone and fitted to a staff. I suppose they were intended to kill seals and fish. They may also kill whales with them as the Esquimaux do. I know not if they are so fond of train oil, but they and everything they have about them smell most intolerable of it.'

'They all retired before dinner and did not wait to partake of our Christmas cheer. Indeed, I believe no one invited them, and for good reason, for their dirty persons and the stench they carried about them was enough to spoil any man's appetite, and that would have been a real disappointment, for we had not experienced such fare for some time. Roast and boiled geese, goose pies &c were victuals little known to us, and we had yet some Madeira wine left, which was the only article of our provisions that was mended by keeping; so that our friends in England did not perhaps celebrate Christmas more cheerfully than we did.'

The humourless and intolerant Forster noted that the crew: 'continued to carouse during two days without intermission, till Captain Cook ordered the greatest part of them to be packed into a boat and put ashore to recover from their drunkenness in the fresh air.'

Remembering his early days as 'a 'prentice boy in the coal trade', Cook thought that 'The SW coast of Tierra del Fuego may be compared to the coast of Norway.' It abounded with wild life. 'It is wonderful to see how the different animals which inhabit this little spot are reconciled to each other; they seem to have entered into a league not to disturb each other's tranquility. The sea lions occupy most of the sea coast, the sea bears take up their abode in the isle; the shags take post on the highest cliffs, the penguins fix their quarters where there is the most communication to and from the sea and the other birds choose more retired places. We have seen all these animals mix together like domesticated cattle and poultry in a farm yard, without one attempting to disturb or molest the other.

Nay, I have often seen the eagles and vultures sitting on the hillocks amongst the shags without the latter, young or old, being disturbed by it. It may be asked how these birds of prey live. I suppose on the carcases of seals and birds which die by various causes, and probably not a few where they are so numerous. This very imperfect account is written more with a view to assist my own memory than to give information to others. I am neither a botanist nor a naturalist and have not words to describe the productions of nature either in the one science or the other.'

Despite his modest appraisal of his own writing, we have found, reading his journals, that his graphic style of descriptive writing succeeds in bringing the place and the moment very close.

'The inner parts of the country were not less savage and horrible. The wild rocks raised their lofty summits till they were lost in the clouds and the valleys lay buried in everlasting snow. Not a tree or shrub was to be seen, no, not even big enough to make a tooth pick.'

'We got on board a little after 12 o'clock with a quantity of seals and penguins, an acceptable present to the crew. It must, however, not be understood that we were in want of provisions; we had yet plenty of every kind . . . but any kind of fresh meat was preferred by most on board to salt. For my own part I was now, for the first time, heartily tired of salt meat of every kind, and preferred the penguins, whose flesh ate nearly as well as bullock's liver. It was, however, fresh and that was sufficient to make it go down.' Cook's capacity for absorbing the coarsest of foods stood him in good stead. The Cleveland farming period of his boyhood had accustomed him to rough, though nourishing, food and laid the foundation of his robust health and vigorous constitution which enabled him to digest everything that came his way.

It was not only salt meat of which he was tired. 'We were in latitude 60° and farther I did not intend to go unless I met with some certain signs of meeting with land, for it would not have been prudent in me to have spent my time in penetrating to the south when it was at least as probable that a large tract of land might be found near Cape Circumcision — besides, I was now tired of these high southern latitudes where nothing was to be found but ice and thick fogs.'

'The risk one runs in exploring a coast in these unknown and

icy seas is so very great that I can be bold to say that no man
will ever venture farther than I have done, and that the lands
which may lie to the south will never be explored. Thick fogs,
snowstorms, intense cold and every other thing that can render
navigation dangerous one has to encounter, and these difficulties
are greatly heightened by the inexpressible horrid aspect of the
country. A country doomed by nature never once to feel the
warmth of the sun's rays, but to lie for ever buried under
everlasting snow and ice.'

'I flatter myself that the intention of the voyage has in every
respect been fully answered, the southern hemisphere sufficiently
explored, and a final end put to the searching after a southern
continent. . . . That there may be a continent or large tract of
land near the Pole I will not deny, on the contrary I am of
opinion that there is, and it is probable that we have seen a
part of it. The excessive cold, the many islands and vast floats
of ice all tend to prove that there must be land to the south,
and that this southern land must lie or extend farthest to the
north opposite the southern Atlantic and Indian oceans.' It has
taken two hundred years for Cook's surmise to be proved
correct without doubt. Only comparatively recently has the
correct outline of the Antarctic land been ascertained. A glance
at an up to date map shows us that the greatest tract of land
there does indeed lie where he said it would.

A seemingly small thing happened at this time, but one on
which the success of the entire voyage almost depended. Their
supply of 'sour krout, a most necessary and valuable article, was
all expended.' And so, although 'my people were yet healthy
and would have cheerfully have gone wherever I thought proper
to lead them, I dreaded the scurvy laying hold of them at a time
when we had nothing left to remove it. Besides it would have
been cruel in me to have continued the fatigues and hardships
they were continually exposed to longer than absolutely
necessary. Their behaviour throughout the whole voyage merited
every indulgence which was in my power to give them.
Animated by the conduct of the officers, they showed
themselves capable of surmounting every difficulty and danger
which came in their way and never once looked upon either the
one or the other to be a bit heightened by being separated from
our companion, the *Adventure*.'

By now the 'sails and rigging were so much worn that something was giving way every hour and we had nothing left

either to repair or replace them. We had been a long time without refreshments, our provisions were in a state of decay and little more nourishment remained in them than just to keep life and soul together.' And so Cook determined to 'steer for the Cape of Good Hope.' A few days before reaching it they met with 'The *True Briton*, Captain Broadly, from China' who 'with a generosity peculiar to the Commanders of the India companies' ships, sent us fresh provisions, tea and other articles which were very acceptable and deserves from me this public acknowledgement. We also got from them a parcel of old newspapers, which were new to us and gave us some amusement in reading.'

At the Cape Cook found a letter from Captain Furneaux awaiting him. He now learnt the reason why the *Adventure* had been unable to follow the *Resolution*. The New Zealanders had murdered and eaten 'ten of his best men in Queen Charlotte's Sound. This, together with a great part of his bread being damaged, was the reason he could not follow me in the route I had proposed to take.'

After taking fresh stores on board they 'now went to work to repair our defects. . . . That our rigging, sails &c should be worn out will not be wondered at when it is known that during this circumnavigation of the globe we have sailed no less than twenty thousand leagues, a distance I will be bold to say was never sailed by any ship in the same space of time before.'

It was here that Cook first saw the published edition of his first *Voyage,* which had been prepared by Hawkesworth, who had seen fit to dress up and elaborate Cook's original manuscript until it was well nigh unrecognisable. He had passed off his own comments and those from other journals as if they were Cook's own words, which they certainly were not. At St Helena, where the *Resolution* called on her way home, the *Journal* had 'given offence to all the principal inhabitants.' Among other things, Hawkesworth had said that the island had no wheel barrows. The islanders made sure Cook knew of this particular inaccuracy, for they studiously placed several outside his lodgings every day.

The Governor of St Helena was none other than John Skottowe, the son of Cook's first patron at Great Ayton. He and his lady, who was a 'very accomplished woman and a native of the island' gave Cook a very pressing invitation to stay with them while he was at the island and they also gave him 'the use of a horse to ride out whenever I thought proper.'

They spent a week at St Helena and were continually fêted.

They 'dined with the Governor, where such an elegant dinner was served up as surprised me, who had never seen more of the island than the barren rocks which compose its borders.' Nor was the ship's company forgotten for 'the second day after our arrival the whole company was entertained at the Governor's Country House, which is situated in a vale, with the sea in front, which to people who are confined the greatest part of the year to the sea shore can be no great addition to the prospect.' He admired the garden where there 'was a live oak tree which seemed to like its situation, and some of the finest peaches I ever tasted.'

'The two preceding evenings before we sailed Mr Graham and Mr Laurel gave each of them a ball. It is to these gentlemen we are obliged for a sight of the celebrated beauties of St Helena, and I should not do my countrywomen at this island justice if I did not confirm the report of common fame; they have fine persons, an easy and genteel deportment and a bloom of colour unusual in a hot climate.'

From there, Cook 'steered for Ascension, where it was necessary for me to touch to take in turtle for the refreshment of my people as the salt provisions they had to eat was what had been in the ship the voyage.'

On the 11th June 1775 they crossed the equator and a month later they reached Fayal in the Azores. 'The chief town is called Villa de Horta and this little city, like all others belonging to the Portuguese, is crowded with Religious Buildings; here are no less than three monasteries of men and two of women, and eight churches.' Cook makes a few notes of the produce available to ships, 'but the entering into the particular productions of each isle is more than I intend, because a better account of these matters may be had, I presume, any day in London from the English merchants who have resided upon them, better than any I can give.'

After leaving these islands Captain Cook 'made the best of my way to England. On Saturday the 29th July 1775 we made the land near Plymouth and the next morning anchored at Spithead. Having been absent from England three years and eighteen days, in which time I lost but four men, and only one of them by sickness.'

'Whatever may be the public judgement about other matters, it is with real satisfaction, and without claiming any merit but of attention to my duty, that I can conclude this account with an observation that our having discovered the possibility of preserving health amongst a numerous ship's company, for such a length of time, in such varieties of climate, and amid such continued hardships and fatigues, will make this voyage remarkable in the opinion of every benevolent person, when the dispute about a southern continent shall have ceased to engage the attention, and to divide the judgement of philosophers.'

CHAPTER 11

THE THIRD VOYAGE

'Such risks as this are the unavoidable companions of
the man who goes on Discoveries.'

After the cannibal incident in New Zealand, Captain Furneaux
brought the *Adventure* to England and arrived there a year
before Captain Cook. He brought with him the Tahitan native,
Omai. Joseph Banks, ever the kindly patron of all travellers
and strangers from distant parts, 'introduced him to a succession
of pleasures.' He was presented at court; he met the Duchess of
Gloucester, who, not having any particular present for him, gave
him her handkerchief, embroidered with a coronet. 'Omai,
gratefully receiving it, immediately kissed the coronet and made
a most respectful bow to the Duchess. This politeness, so
unexpected, gained him the good graces of all present,' George
Young tells us.

'He was fond of intercourse with the great and showed no
inclination for low company or mean habits. He assumed the
dress of an English gentleman, wearing a reddish brown coat
and small clothes, and a white vest and having his hair clubbed
behind. He handled his knife and fork properly at table, his
manners were easy and polite, and he generally acquitted himself
well on his introduction to persons of rank.'

Lord Sandwich, Cook's friend and patron, also undertook to
show the native something of England. He entertained him at
his country seat at Hinchinbroke in Huntingdonshire and took
him to Cambridge, where Omai, himself wearing a military
uniform, was 'much struck with the dress of the doctors and
professors in their robes.' He later accompanied Lord Sandwich
when he reviewed the fleet.

From Joseph Banks's country seat in Lincolnshire, Omai went
with him to the races at York. Here they 'met with the Hon.
Captain Phipps and his brother, the Hon. Augustus Phipps.'
After the races the whole party set out in Mr Banks's carriage
for Mulgrave Castle which lies just outside Whitby. They went
by way of Scarborough, where Omai demonstrated his amazing

swimming ability by taking a young boy of the party, George Colman, out to sea on his back. The heavy coach was all but submerged when it travelled along the sands from Whitby to Mulgrave.

It was the shooting season and Omai was also given a gun. He, however, using South Sea logic, found it easier to take pot shots at the domestic poultry and the Mulgrave chickens were greatly depleted. It is easy to follow his simple reasoning: they went shooting in order to find food. The idea of killing things for 'sport' was quite alien to his primitive mind. He was a sensitive man and could not even bear to see a live worm being used to catch fish.

Captain Cook's return home was a joyful one. Although Mrs Cook had to tell him that the baby, George, had lived only a few months, this homecoming was not marred as the first had been by the sadness of their little daughter's recent death. This time he was greeted by the two boys, who were ten and eleven, both anxious to follow their father in the Navy. Before he left on his third voyage another boy joined them, for Hugh, – named after Sir Hugh Palliser – was born in May 1776. From their Mile End home Cook wrote to his friend John Walker in Whitby:

19 August 1775

Dear Sir,
        As I have not now time to draw up an account of such occurrences of the voyage as I wish to communicate to you, I can only thank you for your obliging letter and kind enquiries after me during my absence. I must, however, tell you that the *Resolution* was found to answer, on all occasions, even beyond my expectations, and is so little injured by the voyage that she will soon be sent out again. But I shall not command her; my fate drives me from one extreme to another: a few months ago the whole southern hemisphere was hardly big enough for me, and now I am going to be confined within the limits of Greenwich Hospital, which are far too small for an active mind like mine. I must, however, confess it is a fine retreat and a pretty income; but whether I can bring myself to like ease and retirement, time will

show.
     Mrs Cook joins me in best respects to you and all your family; and believe me to be, with great esteem,
          Dr. Sir, Your most affectionate friend, and humble Servt., JAMS COOK

This was the time when Cook had his portrait painted by Nathaniel Dance. The work was commissioned by Sir Joseph Banks.

Six months later, another letter to John Walker contained both the momentous news of yet another voyage and an ominous foreshadowing of his not returning from it:

Mile End, 14th February 1776

Dear Sir,
     I should have answered your last favour sooner, but waited to know whether I should go to Greenwich Hospital or the South Sea. The latter is now fixed upon. I expect to be ready to sail about the latter end of April, with my old ship the *Resolution*, and the *Discovery*, the ship lately purchased of Mr Herbert. I know not what your opinion may be on this step I have taken. It is certain I have quitted an easy retirement for an active, and perhaps dangerous voyage. My present disposition is more favourable to the latter than the former; and I embark on as fair a prospect as I can wish. If I am fortunate enough to get safe home, there is no doubt but it will be greatly to my advantage.
     My best respects to all your family; and if any of them come this way I shall be glad to see them at Mile End, where they will meet with a hearty welcome from, Dear Sir, Your most sincere friend,
          and humble servant,
               JAMS COOK

Cook was elected as a Fellow of the Royal Society and his paper on preserving the life of seamen was read to the Society in March 1776 and for this he was awarded the Copley Medal. He had already sailed when the presentation was made and Mrs

Cook received it on his behalf. In a letter to Joseph Banks just prior to his departure he mentions that 'the Council of the Royal Society have decreed me the Prize Medal of this year. I am obliged to you and my other good friends for this unmerited honour.' The medal was given to Mrs Cook by Sir John Pringle, P.R.S., who, Young tells us, 'delivered a judicious and eloquent discourse on the value of Cook's services as a navigator, and particularly of the means which he used for preserving the health of his crew.' Sir John said that they were assembled 'to crown that paper of the year which should contain the most useful and most successful experimental enquiry. Now what enquiry can be so useful as that which hath for its object the saving of the lives of men? And when shall we find one more successful than that before us?'

Part of his discourse referred to Cook's method of obtaining fresh water. 'Not satisfied with plenty, he would have the purest; and therefore whenever opportunity offered he emptied what had taken in but a few days before and filled his casks anew. But was he not above four months in his passage from the Cape of Good Hope to New Zealand, in the frozen zone of the south, without once seeing land? And did he not actually complete his circumnavigation in that high latitude without the benefit of a single fountain? Here was indeed a *wonder of the deep:* Those very shoals, fields and floating mountains of ice among which he steered his perilous course, and which presented such terrifying prospects of destruction; those, I say, were the very means of his support, by supplying him abundantly with what he most wanted.'

Turning to Mrs Cook he said; 'Allow me then, Gentlemen, to deliver this medal, with his unperishing name engraven upon it, into the hands of one who will be happy to receive that trust and to know that this respectable body never more cordially, nor more meritoriously bestowed that faithful symbol of their esteem and affection.'

In February 1776 Captain Cook 'received a commission to command His Majesty's Sloop, the *Resolution.*' Another Whitby ship, the *Diligence* from Langborn's yard, was chosen to accompany him and re-named the *Discovery.* Its command was given to Captain Clerke 'who was my second lieutenant in the *Resolution* last voyage.' By June the ships were ready and they were equipped and 'took on board the necessary stores and provisions for the Voyage, which was as much as we could stow and the best of every kind that could be got.' This included 'a bull, two cows with their calves & some sheep to carry to Tahiti, with a quantity of hay and corn for their subsistence. These were put on board at his Majesty's command and expense, with a view of stocking Tahiti and the neighbouring islands.'

Cook also received 'several astronomical & nautical instruments which the Board of Longitude entrusted to me and Mr King, my second lieutenant; we having engaged to that board to make all the necessary astronomical and nautical observations that should acrue and to supply the place of an Astronomer, which was intended to be sent out in the Ship.'

One of the last letters Cook wrote before he sailed was to his friend Commodore Wilson at Ayton:

> I am at last upon the very point of setting out to join the Resolution at the Nore, and proceed on my voyage, the destination of which you have pretty well conjectured. If I am not so fortunate as to make my passage home by the North Pole, I hope at least to determine whether it is practicable or not. From what we yet know, the attempt must be hazardous, and must be made with great caution. . . . The Journal of my late Voyage will be published in the course of next winter. . . . As to the Journal, it must speak for itself. I can only say that it is my own narrative, and as it was written during the voyage.

One of the primary reasons for the voyage was to search for the North West passage; another was to return Omai to his native country. Cook observed that he left England 'with a mixture of regret and joy. In speaking of England and such persons as had honoured him with their protection and friendship, he would be very low spirited and with difficulty refrain from tears; but turn the conversation to his native country and his eyes would sparkle with joy.'

At Plymouth Cook received on board a party of marines and so 'the overplus men which this reinforcement occasioned were discharged into the ocean.' On Friday, 12th July, 1776 they 'weighed and stood out of the Sound with a gentle breeze at NWBN.' They were off.

Finding the fodder unlikely to last as far as the Cape, he decided to put in at Tenerife, where the Governor, 'sent an

officer on board to compliment me on my arrival.' Cook's fame
had spread all over the globe, and he was given every facility.

As they sailed on, they had 'the mortification to find the
ship exceeding leaky in all her upper works. The hot and dry
weather we had just passed through had opened her seams,
which had been badly caulked at first, so wide that they
admitted the rain water through and there was hardly a man
that could lie dry in his bed. ...... The sails in the sail room got
wet and before we had weather to dry them many of them were
quite ruined. ... This complaint of our sail rooms we
experienced on my late voyage, and was represented to the yard
officers who undertook to remove it, but it did not appear to
me that anything had been done that could answer that end. To
repair these defects, the caulkers were set to work as soon as
we got into fair settled weather, to caulk the decks and inside
weather works of the ship, for I would not trust them over the
side while at sea.'

They 'proceeded on our voyage without meeting with any
thing of note' until the middle of October, when they arrived
at the Cape of Good Hope. There, they heard that a French
ship had recently been wrecked and plundered. The Dutch, 'by
way of excusing themselves from being guilty of a crime that
is a disgrace to every civilised state, endeavoured to lay the
whole blame on the Captain for not applying for a guard in
time. Without mentioning obstacles that were thrown in his
way when he did apply, if we may believe the account above
mentioned, when the guard did come things were not a bit
mended but rather worse. In short, the Dutch in this affair
strictly adhered to the maxim they have laid down at this
place which is, to get as much by strangers as they possibly can
without ever considering whether the means are justifiable or
not.'

Cook himself soon experienced this, for, despite his care in
penning up the sheep at night, they put a dog in their midst
and stole most of them. 'They tell us that the police are so well
executed here that it is hardly possible for a slave with all his
cunning and knowledge of the country to escape, yet my sheep
evaded the vigilance of all the fiscal's officers and people. However, after some trouble and expense in employing some of the
meanest and lowest scoundrels in the place, who — to use the
phrase of the person who recommended me to this method —
would for a ducatoon cut their master's throat, burn the house

over his head and bury him and the whole family in the ashes, I
recovered them all but the two ewes.' He was offered two by
the Second Governor who had 'taken some pains to introduce
European sheep at the Cape, but his endeavours have been
frustrated by the obstinacy of the country people, who value
their own sheep better.' Cook concludes the account by
noting: 'Were it not for the slaves that are continually
importing here, the Dutch settlement at the Cape would be
thinner of inhabitants than any habitable part of the world
whatever.'

'After the disaster which happened to our sheep, it may well
be supposed I did not let those that remained stay long on
shore, but got them and the other cattle on board immediately.
To which I added two young bulls, two heifers, two young
horses, two mares, two rams, several ewes and goats and some
rabbits and poultry, all of them intended for New Zealand,
Tahiti and the neighbouring islands, or any other place we
might meet, where there was a prospect that the leaving of
them might prove useful to posterity.' They took provisions
aboard and 'every other necessary thing we could think of for
such a voyage, neither knowing when nor where we should come
to a place where we could supply ourselves so well.'

This veritable Noah's Ark, at the beginning of December 1776,
'weighed and put to sea with a light breeze at south.' They
soon ran into 'a very high sea which made the ship roll and
tumble exceedingly and gave us a deal of trouble to preserve
the cattle we had on board, and notwithstanding all our care,
several goats and some sheep died, owing in great measure to
the cold, which we began now most sensibly to feel.'

Christmas saw them at Kerguelen Island. Cook immediately
set them to work collecting water and on the 27th, 'the people
having worked hard the preceding day and nearly completed our
water, I gave them this to celebrate Christmas. . . . Here I
displayed the British flag and named the harbour Christmas
Harbour as we entered it on that festival.' The tops of all the
hills were 'capped with snow and they appeared to be full as
naked and barren as any we had seen. Not the least sign of a
tree or shrub was to be seen either inland or on the coast.'

'After leaving the island of desolation I steered EBN, intending
to touch at New Zealand to recruit of water, take in wood and
make hay for the cattle. Their number by this time was
considerably reduced, two young bulls and one of the heifers

were dead, as also the two rams and most of the goats.'

They came first to Tasmania. Cook noted two rocks, one of which he named *Eddystone* 'from its very great resemblance to that lighthouse. Nature seems to have left these two rocks here for the same purpose that the Eddystone lighthouse was built by man, viz, to give navigators notice of the dangers about them, for they are the elevated summits of a ledge of rocks under water on which the sea in many places breaks very high; their surface is white with the dung of sea fowl, so that they may be seen some distance, even in the night.' Here they paused long enough to take in wood and water and meet some of the natives who 'came out of the woods to us without showing the least mark of fear and with the greatest confidence imaginable.'

On the 11th January 1777 they reached New Zealand and Cook 'anchored in our old station in Queen Charlotte's Sound.' At first the natives were very reluctant to come aboard for 'they were apprehensive we were come to revenge the death of Captain Furneaux's people.' When they found that Cook intended to continue his previous friendship they 'laid aside all manner of restraint and distrust.'

The watch for scurvy was constant. 'Celery, scurvy grass and portable soup were boiled with the peas and wheat for both ships' companies every day during our whole stay, and they had spruce beer for their drink; so that if any of them had contracted any seeds of the scurvy these articles soon removed it.'

Cook watched the natives build an instant village. 'While the men were employed raising the huts the women were not idle, some were taking care of the canoes, some securing the provisions. . . . others went to gather dry sticks to make a fire to dress their victuals. As to the children, I kept them — as also some of the more aged — sufficiently employed in scrambling for beads till I had emptied my pockets, and then I left them.'

Their 'articles of commerce were Curiosities, Fish and Women. The two first always came to a good market, which the latter did not. The seamen had taken a kind of dislike to these people and were either unwilling or afraid to associate with them. It had a good effect as I never knew a man quit his station to go to their habitation. A connection with women I allow because I cannot prevent it, but never encourage.'

The natives soon overcame their shyness and 'great numbers daily frequented the ships and the encampment on shore. What partly induced them to resort to the latter more than usual was some seal blubber we were melting down. No Greenlander can be fonder of train oil than these people, the very dregs of the casks and skimmings of the kettle they eat, but a little pure oil was a feast they seemed not often to enjoy.'

Two of the chiefs 'begged of me some goats and hogs' and were given two of each. 'They made me a promise not to kill them, but in this I put no great faith.'

Omai 'expressed a desire to take one of the natives with him to his own country' and volunteers were not lacking. 'A youth about 17 or 18 years of age' offered to go and took with him 'a boy about 9 or 10. He was presented to me by his own father with far less indifference than he would have parted with his dog. It was to no purpose my endeavouring to convince these people of the improbability or rather impossibility of these youths ever returning. Not one, even their nearest relations, seemed to trouble themselves about what became of them. Since this was the case I was well satisfied the boys would not be losers by exchange of place, and therefore the more readily gave my consent to their going.'

However, 'we had no sooner lost sight of land than our two adventurers, what from sea sickness and reflection, repented heartily of the step they had taken. All the soothing encouragement we could think of availed but little. They wept both in public and in private and made their lamentation in a kind of song. . . . Thus they continued for many days, till their sea sickness wore off and the tumult of their minds began to subside. The fits of lamentation became less and less frequent and at length quite went off, so that their friends and their native country were no more thought of and they were as firmly attached to us as if they had been born among us.'

They sailed to what are now the Cook Islands and traded with the natives, who brought plantains and cocoanuts for which they 'demanded a dog and refused every other thing that was offered. We had a dog and a bitch on board belonging to one of the gentlemen that were a great nuisance in the ship and might have been disposed of here to some essential purpose, but such a purpose the owner never intended them. However, to gratify these people, Omai parted with a favourite dog he had and they departed highly satisfied.'

Cook always paid for everything he got. Although one small island appeared to be uninhabited, after they had taken cocoanuts and fodder for the cattle, 'Mr Gore found a few empty huts, in one of which he left a hatchet and some nails to the full value of what we took from the island.'

Cook 'determined to bear away for the Friendly Isles, where I was sure of being supplied with everything I wanted.' In order to save water he ordered the still to be brought into use. 'There has lately been made some improvement, as they are pleased to call it, to this machine, which in my opinion is much for the worse.' Later, they had heavy rain and 'finding we could get more by the rain in an hour than by the still in a month, I laid it aside as a thing attended with more trouble than profit.'

They reached the Friendly Islands (Tonga) in May, 1777. They were well received, but found the natives' thieving ability as great as ever. 'These people very frequently took opportunities to show us what expert thieves they were. Even some of the chiefs did not think this profession beneath them. One was caught carrying out of the ship concealed under his clothes a bolt belonging to the winch, for which I ordered him a dozen lashes and made him pay a hog for his liberty; after this we were not troubled with thieves of rank, their servants were employed in this dirty work, on which a flogging made no more impression than it would have done on the main-mast.'

Captain Clerke 'hit upon a method which had some effect. This was by shaving their heads . . . it was looked upon as a mark of infamy and marked out the man. But one or two of our gentlemen, whose heads were not overburdened with hair, lay under violent suspicion.'

The visits to the various islands in the group were enlivened by entertainments put on for their benefit. At one place there were 'wrestling and boxing matches. The first were performed in the same manner as at Tahiti and the second very little different from the method practised in England. But what struck us with the most surprise was to see a couple of lusty wenches step forth and without the least ceremony fall to boxing, and with as much art as the men.'

They vied with each other to provide entertainment. Cook put the Marines through their paces and then the Chief 'entertained us again in his turn with a sight entirely new. It was a kind of dance, performed by men and youths of the first rank; but so much unlike any thing I know of in any part of the world, that no description I can give will convey even a tolerable idea of it.' It was a sort of figure marching display, each man holding 'an instrument something like a paddle, of 2½ feet in length with a small handle and thin blade, so that these were very light and the most of them very neatly made.' It was accompanied 'by a song in which everyone joined as with one voice; it was musical and harmonious and all their motions were performed with such justness that the whole party moved and acted as one man. It so far exceeded anything we had done to amuse them that they seemed to pique themselves on the superiority they had over us.'

After this Cook put on a firework display 'which astonished and pleased them beyond measure and entirely turned the scale in our favour.'

The gentle inhabitants treated the visitors with great generosity. The Chief of Lifuka collected two great piles of yams, bread fruit, plantains, cocoanuts and sugar cane, eight pigs, some fowls and two turtles; one, he said, was for Omai and the other, 'which was about two thirds of the whole, was for me. He told me I might take it on board whenever it was convenient, nor was it necessary to leave anybody to look after it, for not so much as a single cocoanut would be taken away, and so it proved, for I left everything behind me and returned on board to dinner, carrying the chief along with me, and yet when I came to send for them on board nothing was missing. There was as much as loaded four boats and far exceeded any present I ever before received from an Indian Prince. I took the opportunity of his being on board to make him a return so much to his satisfaction, that as soon as he went ashore he made me another present of two large hogs, a quantity of cloth and some yams.'

The King of Tonga was always treated with the greatest respect. His subjects paid him obeisance by 'bowing the head to the sole of his foot and touching or tapping the same with the upper and under sides of the fingers of both hands. It appeared that the king could not refuse anyone who chose to pay him this compliment, for the common people would frequently take it into their heads to do it when he was walking and he was always obliged to stop and hold up one of his feet behind him till they had done. This, to a heavy, unwieldy man like him must be attended with some trouble and pain, and I have seen him make a run, though very unable, to get out of the way or to a place where he could sit down.' Cook was 'quite charmed with the decorum that was observed, I had no where seen the like, no, not even amongst more civilised nations.'

The king frequently dined with Cook. 'It was very convenient for when he was there everyone else was not only excluded
the table, but few would remain in the cabin.' At other times
Cook was importuned so much by the lesser chiefs that he
'could never sit down to a dinner with any satisfaction. The
King was very soon reconciled to our way of cooking, but I
believe he dined with me more for the sake of the drink than
the victuals, for he loved a glass of wine as well as most men
and was as cheerful over it.'

The King returned the hospitality in no uncertain manner.
When Cook went ashore he 'found his people very busy fixing
four very long posts.' The space between the posts was then
filled up with yams and as they went on they fixed sticks across
and continued until each pile was 'the height of 30′ or upwards.
On the top of one they placed two baked hogs, on the top of
the other a living one and another they tied by the legs about
half way up. It was extraordinary to see with what facility and
despatch they raised these piles. Had our seamen been ordered
to do such a thing they would have sworn it could not be done
without carpenters, and the carpenters not without a dozen
different sorts of tools, and the expense of at least a
hundredweight of nails and after all it would have employed
them as many days as it did these people hours. But seamen,
like most other amphibious animals, are always the most helpless
on land.'

The completed piles were decorated with 'different sorts of
small fish. . . . As to the fish, it might serve to please the sight,
but was very offensive to the smell, as some of it had been kept
two or three days for this occasion.'

Going from one island to another, the *Resolution* 'by a
small shift of wind fetched farther to windward than was
expected; by this means she was very near running plump upon
a low sandy isle surrounded by breakers. . . . Such risks as this
are the unavoidable companions of the man who goes on
Discoveries.'

Cook left Cattle and sheep at Tonga and as he walked up a
hill 'whilst I was viewing these delightful spots I could not help
flattering myself with the idea that some future navigator may,
from the very same station, behold these meadows stocked
with cattle the English have planted at these islands.' He also
'planted a pineapple and sowed the seeds of melons &c in the
Chief's plantation, and had a dish of turnips to dinner, being the

produce of the seeds I left last voyage.'

They 'took leave of the Friendly Islands and their inhabitants after a stay of between two and three months, during which time we lived together in the most cordial friendship. Some accidental differences, it is true, now and then happened owing to their great propensity to thieving, but too often encouraged by the negligence of our own people. But these differences were never attended with any fatal consequences, to prevent which all my measures were directed. Also during this time we expended very little of our sea provisions, but lived on the produce of the islands; and besides the opportunity of leaving the cattle before mentioned among them, those designed for Tahiti received fresh strength.'

CHAPTER   12

OMAI'S   RETURN

'I called all hands together and acquainted them
with what further was expected to be done in the
voyage.'

They came to Tahiti in August, 1777. 'Omai's brother in law,
who chanced to be here, came on board and three or four more,
all of whom Omai knew before he embarked with Captain
Furneaux, yet there was nothing either tender or striking in
their meeting. On the contrary, there seemed to be a perfect
indifference on both sides, till Omai asked his brother down
into the cabin, opened the drawer where he kept his red
feathers and gave him a few. This being presently known to those
on deck, the face of affairs was entirely turned. . . . It was
evident to everyone that it was not the man but his property
they were in love with, for had he not shown them his red
feathers, which is the most valuable thing that can be carried to
the island, I question if they had given him a cocoanut.'

At long last Cook was able to deliver the livestock, 'viz: A
peacock and hen which my Lord Besborough was so kind to
send me for this purpose a few days before I left London; a
turkey cock and a hen; one gander and three geese; a drake and
four ducks, . . . three cows, the bull, the horse and mare and
sheep I put ashore at Matavai. And now found myself lightened
of a very heavy burden, the trouble and vexation that attended
the bringing of these animals thus far is hardly to be conceived.
But the satisfaction I felt in having been so fortunate as to fulfil
His Majesty's design in sending such useful animals to two
worthy nations sufficiently recompensed me for the many
anxious hours I had on their account.'

There is an unhappy postscript to this extraordinarily generous
act. Years later William Bligh, Cook's Master on this voyage,
visited Tahiti in the *Bounty* and learnt that all the livestock had
been destroyed by enemy action and not one remained. Omai
fared little better, for Bligh learnt that he lived only two years
after Cook left.

They stayed at Tahiti for two months, refreshing the ships'

companies and putting the vessels into trim in preparation for their haul northwards. Cocoanuts were plentiful and 'thinking it a good time to save our spirits, I called all hands together and acquainted them with what further was expected to be done in the voyage. I pointed out to them the improbability of our getting any supplies of provisions after leaving the Society Isles and the hardship they would feel at being in short allowance in a cold climate, and that rather than run this risk it was better for them to be without grog now, but that I left it entirely to their own choice. I had the satisfaction to find that it remained not under a moment's consideration, but was consented to immediately. The next day I ordered Captain Clerke to make the same proposal to his people, which they also consented to, and we stopped serving grog except on Saturday nights, when they had full allowance to drink to their female friends in England, lest among the pretty girls of Tahiti they should be wholly forgotten.'

The natives were as cunning as ever. Cook tells the story of a certain chief, Otoo, to whom he had given a spy glass; 'After he had had it two or three days, and probably finding it of no use, he carried it privately to Captain Clerke and told him that as he had been his very good friend he had got a present for him which he knew he would like, but, says Otoo, "You must not let Toote (Cook) know it because he wants it and I would not let him have it," and then he puts the glass into Captain Clerke's hand, at the same time assuring him that he came honestly by it. . . . Being willing to oblige Otoo, and thinking that a few axes would be of more use than the glass, he got out four to give him in return, which Otoo no sooner saw than he said: "Toote offered me five for it." "Well," says Captain Clerke, "If that is the case, your friendship shall not make you a loser. There's six for you." These he accepted, but desired again that I might not be told of what he had done.'

They learnt from Otoo that Spaniards had landed at the island 'who were to return and bring with them houses, all kinds of animals and men and women who were to settle, live and die on the island. . . . It was easy to see that the idea pleased him, little thinking that such a step would at once deprive him of his Kingdom and the people of their liberties. This shows with what facility a settlement properly conducted might be made among them, which for the regard I have for them I hope will never happen.'

Cook was unimpressed by a large cross which the Spaniards
had erected with *Christus Vincit Carolus III Imperat 1774* on
it. 'On the other side of the post which supported the cross I
had cut out, *Georgius tertius Rex Annis 1767, 69, 73, 74 & 77.*'

It was decided to leave Omai and the two New Zealand boys
at Huahine, for 'after he got clear of the gang that surrounded
him at Tahiti, he behaved with such prudence as to gain respect,'
and to this end 'the carpenters of both ships were set to work
to build a small house for Omai.' At the same time 'some hands
were employed making a garden, planting shaddocks, vines,
pineapples, melons and several other articles.' The house was
built 'of boards and with as few nails as possible, that there
might be no inducement to pull it down.' For 'here as in many
other countries, a man that is richer than his neighbours is sure
to be envied.'

While this was being done they 'got the bread remaining in
the bread room ashore to clear it of vermin. The number of
cockroaches that were in the ship at this time is incredible.
The damage they did us was very considerable and every method
we took to destroy them proved ineffectual.' There were rats
too. 'The ship being a good deal pestered with rats, I hauled
her within thirty yards of the shore and made conveniences for
them to go shore, being in hopes some would be induced to
it, but I believe we got clear of very few, if any.'

Cook was bothered with rheumatism. Hearing of this, the
chief's sister and about a dozen other women came to the
Resolution. 'They told me they were come to sleep on board
and to cure me of the disorder that I complained of, which
was a sort of rheumatic pain in one side from the hip to the
foot. This kind offer I accepted of, made them up a bed in the
cabin floor and submitted myself to their direction. I was desired
to lay down in the midst of them, and as many as could get
round me began to squeeze me with both hands from head to
foot, but more especially the parts where the pain was, till they
made my bones crack and a perfect Mummy of my flesh — in
short, after being in their hands about a quarter of an hour I was
glad to get away from them. However, I found immediate relief
from the operation. They gave me another rubbing down before
I went to bed and I found myself pretty easy all the night after.
They repeated the operation the next morning and again in the
evening, after which I found the pains entirely removed. This
they call *Romy*, an operation which in my opinion far exceeds

the flesh brush or anything we make use of of the kind. It is universally practised among them, it is some times performed by the men, but more generally by the women. If at any time one appears languid or tired and sits down by any one of them, they immediately begin with Romy upon your legs, which I have always found to have an exceeding good effect.'

The time for Omai's departure came and, with the two New Zealand youths, he went ashore 'after taking a very affectionate farewell of all the officers. He sustained himself with a manly resolution till he came to me, then his utmost efforts to conceal his tears failed, and Mr King, who went in the boat, told me he wept all the time in going ashore. Whatever faults this Indian had they were more than balanced by his great good nature and docile disposition. During the whole time he was with me I seldom had reason to find fault with his conduct. His grateful heart always retained the highest sense of the favours he received in England, nor will he ever forget those who honoured him with their protection and friendship during his stay there.

'He was not a man of much observation. There were many little arts as well as amusements among the Friendly Islands which he might have conveyed to his own, but I never found that he used the least endeavour to make himself master of any one. This kind of indifferency is the true character of his nation. Europeans have visited them for these ten years past, yet we find neither new arts nor improvements in the old, nor have they copied after us in any one thing. We are therefore not to expect that Omai will be able to introduce many of our arts and customs amongst them. I think, however, he will endeavour to bring to perfection the fruits &c we planted, which will be no small acquisition.'

Two men deserted. 'As these were not the only two persons in the ships who wanted to end their days at these islands, it was necessary in order to put a stop to further desertion to have them got back at all events.' The story of their recapture is worthy of the best of present day Hollywood film epics. Cook determined to take hostages in their place. 'The Chief, his son, daughter and son-in-law came on board. The three last I resolved to detain till they brought back the two deserters. With this view, Captain Clerke invited them on board his ship and secured them in his cabin.' This caused great consternation and there were 'great lamentations under the ship's stern all day long. The old women cut their heads till the blood runs down in streams on their shoulders, which they smear about their bodies and make a terrible howling.' Then all at once, 'all the Indians that were in and about the harbour and ships began to move off as if some sudden panic had seized them.' Which it had, for they heard that Captain Clerke and Mr Gore had been captured ashore. 'There was no time to deliberate, I instantly ordered the people to arm, and in less than five minutes a strong party under the command of Mr King was sent to rescue them; at the same time two armed boats were sent after the flying canoes and a party under Mr Williamson to cut off their retreat to the shore. These several detachments were hardly out of sight before an account arrived that we had been misinformed.'

It was Cook himself they were after. 'Their first and greatest desire was to have got me. It was my custom to go and bathe in the fresh water every evening, and very often alone and always without arms; expecting me to go as usual this evening, they had determined to seize me, and Captain Clerke too if he had accompanied me. But I had, after confining their people, determined not to put myself in their power and had cautioned both Captain Clerke and the officers not to go far from the ships. The chief asked me three times if I would not go to the water.'

'The affair was first discovered by a girl which one of the officers brought from Huahine, she overhearing some of her countrymen say they would seize Captain Clerke and Mr Gore, ran to acquaint the first of our people she met with. Those who were charged with the execution of the design threatened to kill her as soon as we were gone for disappointing them, so that her friends came some days after in the middle of the night and took her out of the ship to convey her to a place of safety till they had an opportunity to send her home.'

The deserters were recovered and duly punished and the hostages returned to their country.

It was almost inevitable that the island which they discovered on 24th December 1777 should be called Christmas Island. It was found to be rich in turtles, but 'not a drop of fresh water was found on the whole island.' Two seamen managed to lose themselves for twenty-four hours. 'It was a matter of surprise to everyone how these men contrived to lose themselves. The land over which they had to travel from the sea coast to the lagoon was not more than three miles across, . . . from many parts of which the ship's masts were to be seen: but this was a thing they never once thought of looking for, nor did they know in what direction the ships were from them, nor which way to go to find either them or the party, no more than if they had just dropped from the clouds. Considering what a strange set of beings the generality of seamen are when on shore, instead of being surprised at these men losing themselves, we ought rather to have been surprised there were no more of them.'

In January 1778 'At daybreak on the morning of the 18th an island was discovered bearing NEBE and soon after we saw more land bearing north.' This was Kauai, in the Hawaiian Islands. 'We were in some doubt whether the land was inhabited. This doubt was soon cleared up by seeing some canoes coming off from the shore towards the ship. I immediately brought to to give them time to come up. There were three and four men in each and we were agreeably surprised to find them of the same nation as the people of Tahiti and the other islands we had lately visited. It required but very little address to get them to come alongside, but we could not prevail on any to come on board. They exchanged a few fish they had in the canoes for anything we offered them, but valued nails or iron above every other thing.'

The next morning they 'stood in for the land and were met by several canoes filled with people, some of them took courage and ventured on board. I never saw Indians so much astonished at the entering a ship before; their eyes were continually flying from object to object. The wildness of their looks and actions fully expressed their surprise and astonishment at the several new objects before them and evinced that they had never been on board of a ship before.'

Cook sent 'three armed boats under the command of Lieutenant Williamson, to look for a landing place and fresh water.' He attempted to land in one place, but was 'prevented by the Indians coming down to the boat in great numbers, and were for taking away the oars, muskets, and in short, everything they could lay hold on and pressed so thick upon him that he was obliged to fire, by which one man was killed. But this unhappy circumstance I did not know until after we left the islands, so that all my measures were directed as if nothing of the kind had happened.'

These measures included orders that 'no women, on any account whatever, were to be admitted on board the ships. I also forbade all manner of connection with them and ordered that none who had the veneral upon them should go out of the ships. But whether these regulations had the desired effect or no, time can only discover. It is no more than what I did when I first visited the Friendly Islands, yet I afterwards found it did not succeed, and I am much afraid this will always be the case where it is necessary to have a number of people on shore. . . . It is also a doubt with me that the most skilful of the Faculty can tell whether every man who has the veneral is so far cured as not to communicate it further.'

The ships anchored at the island of Kauai and Cook 'went ashore with three boats to look at the water and try the disposition of the inhabitants, several hundreds of whom were assembled on a sandy beach before the village. The very instant I leaped ashore they all fell flat on their faces and remained in that humble posture till I made signs to them to rise.' They traded for 'hogs and potatoes, which the people gave us in exchange for nails and pieces of iron formed into something like chisels. We met with no obstruction in watering, on the contrary, the natives assisted our people to roll the casks to and from the pond. As soon as everything was settled to my satisfaction, I left the command to Mr Williamson, who was with me, and took a walk up the valley accompanied by Dr Anderson and Mr Webber; conducted by one of the natives and attended by a tolerable train. Our guide proclaimed our approach and everyone whom we met fell flat on their faces and remained in that position till we had passed. This, as I afterwards understood, is done to their great chiefs.' They saw some graves and the native guides 'gave us clearly to understand that three human sacrifices had been buried there.'

'The ground over which I walked was in a state of nature, very stony, and the soil seemed poor; it was, however, covered with shrubs and plants, some of which sent forth the most fragrant smell I had anywhere met with in this sea.'

Pausing only long enough to take in fresh supplies, Cook named the group *Sandwich Islands* 'in honour of the Earl of Sandwich', and sailed northward at the beginning of February 1778. He noted in his journal that they were 'an open, candid, active people and the most expert swimmers we had met with; in which they are taught from their very birth. It was very common for women with infants at the breast to come off in canoes to look at the ships, and when the surf was so high that they could not land them in the canoe, they used to leap

overboard with the child in their arms and make their way
ashore through a surf that looked dreadful.'

CHAPTER 13

AMERICA

'There can be little doubt but there is a northern communication of some sort between this ocean and Baffin's Bay, but it may be effectually shut up against shipping by ice.'

For a month they sailed northwards and on 6th March, 1779 they reached the north western coast of America. 'Saw two seals and several whales and at day break the next morning the long looked for coast of New Albion was seen. . . . The land appeared to be of a moderate height, diversified with hill and valley, and almost-everywhere covered with wood. . . . At the northern extreme the land formed a point, which I called *Cape Foul Weather*, from the very bad weather we soon after met with.' This occasioned them to tack, off and on, for some time, but eventually they were able to sail towards the land once more. Cook saw *'Cape Blanco,* discovered, or seen by Martin de Agualar the 19th of January 1603. It is worth observing that in the very latitude we were now in, Geographers have placed a large entrance or Strait, the discovery of which they ascribe to the same Captain, whereas nothing more is mentioned in the Voyage than seeing a large river, which he would have entered but was prevented by the currents.'

Standing north, 'there appeared to be a small opening in the land which flattered us with hopes of finding a harbour. These hopes lessened as we drew nearer, and at last we had some reason to think that this opening was closed by low land. On this account I called the point of land to the north of it *Cape Flattery.'*

The severe winds drove him off shore, nevertheless 'it was by the means of these southerly blasts that we got to the NW at all.' They came to a point where 'the shore forms a large bay, which I called *Hope Bay*, in which from the appearance of the land, we hoped to find a good harbour and the event proved we were not mistaken.' As they drew near, 'we found the coast to be inhabited and the people came off to the ships in canoes without showing the least mark of fear or distrust. We had at one time thirty-two canoes filled with people about us. They seemed to be a mild, inoffensive people, showed great readiness to part with anything they had and took whatever was offered them in exchange.'

They moored the following day and 'as we found the ship very leaky in her upper works, the caulkers were set to work.' They 'had the company of the natives all the day, who now laid aside all manner of restraint, if they ever had any, and came on board the ships and mixed with our people with the greatest freedom. And we soon found they were as light fingered as any people we had before met with, and were far more dangerous, for with their knives and other cutting instruments of iron, they would cut a hook from a tackle, or any other piece of iron from a rope, the instant our backs were turned. We lost the Fish hook, a large hook between 20 and 30 pounds weight, several lesser hooks and other articles of iron in this manner, and as to our boats, they stripped them of every article of iron about them worth carrying away, though we had always men in them to guard them, but one fellow would amuse the boat keeper at one end while another was pulling her to pieces at the other.'

The natives brought off 'skins, curiosities &c, and such was the passion for these things among our people that they always came to a good market, whether they were of any value or no. ... Nothing would go down with them but metal, and brass was their favourite. So that before we left the place, hardly a bit of brass was left in the ship, except what was in the necessary instruments. Whole suits of clothes were stripped of every button, bureaus &c of their furniture, and copper kettle, tin cannisters, candlesticks &c all went to wreck.'

In Nootka Sound he went ashore to the Indian village of Yuquot, where the people 'who were numerous and to most of whom I was known, received me very courteously, every one pressing me to go into his house, or rather apartment, for several families live under the same roof, and there spread a mat for me to sit down upon and showed me every other mark of civility.' Nevertheless Cook wrote of them: 'They are full as dirty in their victuals and cookery as in their persons. Their houses are as filthy as hog-sties, every thing in them stinks of fish, train oil and smoke.'

'Having a few goats and two or three sheep left,' Cook went

to find some grass for them, not thinking the natives would mind, 'but it proved otherwise, for the moment our people began to cut they stopped them' and demanded payment. 'There was not a blade of grass that had not a separate owner, so that I very soon emptied my pockets.'

'Of the vegetables this place produceth we benefitted by none except the spruce tree of which we made beer, and wild garlic' which the natives brought in profusion. It later transpired that the seamen resolved not to drink this spruce beer, but fortunately for them 'they did not attempt to carry their resolution into execution' and it was only by chance that Cook came to hear of it.

On 27th April 1778 they 'stretched off to the SW under all the sail the ships could bear . . . at daylight the next morning we were quite clear of the coast. The *Discovery* being some distance astern, I brought to till she came up and then bore away and steered NW, the direction I supposed the coast to take.' Once again he was right.

'The *Resolution* sprung a leak which at first alarmed us not a little. It was found to be under the starboard buttock, where from the bread room we could both hear and see the water rush in . . . the fish room was found to be full of water and the casks in it afloat. . . . After the water was bailed out, which employed us till midnight, one pump kept it under, which gave us no small satisfaction.' All their 'endeavours to stop the leak at sea proving ineffectual', Cook determined to find a safe anchorage where the ship could be properly mended. Eventually they moored 'in a place which in the chart is distinguished by the name of *Snug Corner Bay*, and a very snug place it is . . . sheltered from all winds.' By 17th May 1778, 'the leak being stopped and the sheathing made good, at 4 o'clock in the morning we weighed and steered to the NW.'

Ten days later, sailing up the unknown, fog-bound coast, sounding, charting and naming as he went, he saw 'a very lofty promontory whose elevated summit forming two exceeding high mountains was seen above the clouds. This promontory I named *Cape Douglas* in honour of my very good friend, Dr Douglas, Canon of Windsor.' It was Douglas with whom Cook had left the manuscript of his second voyage to prepare for publication; he also edited the *Journal* of this third voyage.

It must be remembered that the main purpose of this voyage (after returning Omai to his homeland) was to determine the existence of a North West Passage. 'It was imagined that the land on our larboard, north of Cape Douglas, was composed of a group of islands, disjoined by so many channels, any one of which we might make use of according as the wind served. With these ideas, having a fresh gale at NNE we stood to the NW till 8 o'clock, when we clearly saw that what we had taken for islands were the summits of mountains that were everywhere connected by lower land which the haziness of the horizon had prevented us from seeing at a greater distance. This land was everywhere covered with snow from the summits of the hills down to the very sea beach, and had every other appearance of being part of a great Continent, so that I was fully persuaded that we should find no passage by this inlet and my persevering in it was more to satisfy other people than to confirm my own opinion.'

As they travelled on the air was 'raw and cold' and for days at a time they had 'almost constant misty weather'. There was good reason for calling one place *Foggy Cape*. They daily 'saw most of the sea birds that are commonly found in most other northern oceans, such as gulls, shags, puffins, sheer-waters &c and sometimes ducks, geese and swans, and seldom a day passed without seeing seals, whales and other large fish.' Cook thought that this country was 'more broken or rugged than any part we had yet seen. ... Every part had a very barren appearance and was covered with snow from the summits of the highest hills down to a very small distance from the sea coast.' They saw a volcano, 'which continually threw out a vast column of black smoke. It stands not far from the coast and in the latitude of 54° 48′ N. It is also remarkable from its figure, which is at the very summit. We seldom saw this or any of the other mountains wholly clear of clouds, at times both base and summit would be clear, when a narrow cloud, sometimes two or three one above the other, would embrace the middle like a girdle which, with the column of smoke rising perpendicular to a great height and spreading before the wind into a tail of vast length, made a very picturesque appearance.'

In one place they were navigating in dense fog and 'hearing the sound of breakers on our larboard bow' they immediately anchored. 'A few hours after the fog cleared away a little and it was perceived we had escaped very imminent danger' for there were two elevated rocks on either side of them. 'There were several breakers about them and yet Providence had

conducted us through between these rocks where I should not
have ventured in a clear day, and to such an anchoring place
that I could not have chosen a better.'

On the 3rd August, 1778, 'Mr Anderson, my Surgeon, who
had been lingering under a consumption for more than twelve
months, expired between 3 and 4 this afternoon. He was a
sensible young man, an agreeable companion, well skilled in his
profession, and had acquired much knowledge in other sciences,
that had it pleased God to have spared his life, might have been
useful in the course of the voyage. Soon after land was seen to
the westward, it was supposed to be an island and to perpetuate
the memory of the deceased, for whom I had a very great
regard, I named it Anderson's Island.'

They landed on another small island where they found 'wild
parsley, pease, longwort &c, some of which we took on board
for the pot.' They also found a sledge and so named it Sledge
Island.

Up in Bering Strait, on the Asian side, they found the Chucchi
people to be more akin to Esquimaux than the others they had
encountered. 'All the Americans we had seen before were rather
low of stature with round chubby faces and high cheekbones,
whereas these are long visaged, stout made men and appeared
to be quite a different nation.' 'They seemed very fearful and
cautious, making signs for no more of our people to come up,
and on my laying my hand on one man's shoulder he started
back several paces.'

Back on the American coastline 'where the coast here forms
a point named Point Mulgrave the land appeared very low next
the sea but a little back it rises into hills of a middling height,
the whole was free from snow and to appearance destitute of
wood.' Unfortunately this name, given to commemorate the
second Lord Mulgrave, has not survived.

In the middle of August they 'perceived a brightness in the
northern horizon like that reflected from ice, commonly called
the blink. It was little noticed from a supposition that it was
improbable that we should meet with ice so soon, and yet the
sharpness of the air and gloomyness of the weather for two or
three days past seemed to indicate some sudden change. At 1 pm
the sight of a large field of ice left us no longer in doubt
about the cause of the brightness of the horizon.'

After a long period of beating about and searching for an
entrance through the ice, Cook 'gave up the design I had of

plying to the westward and stood off shore again. The season
was now so far advanced and the time when the frost is expected
to set in so near at hand that I did not think it consistent with
prudence to make any further attempt to find a passage this
year in any direction, so little was the prospect of succeeding.

'My attention was now directed towards finding out some
place where we could wood and water, and in the considering
how I should spend the winter, so as to make some improvement
to Geography and Navigation and at the same time be in a
condition to return to the north in further search of a passage
the ensuing summer.'

The frost did indeed come: 'For these two days past the mean
height of the mercury in the thermometer has been very little
above the freezing point and often below it, so that the water
in the water vessels on deck was frequently covered with a
sheet of ice.'

After sighting the coast of Asia and admiring Behring's
accuracy — 'In justice to Behring's memory I must say he has
delineated this coast very well and fixed the latitude and
longitude of the points better than could be expected from
the methods he had to go by,' Cook returned to Sledge Island
in order to get some wood 'which we began to be in great need
of.'

That there was always ice in this part of the globe, winter
and summer 'none who has been upon the spot will deny, and
none but closet studying philosophers will dispute.'

In Norton Sound 'a man came off in a canoe, I gave him a
knife and a few beads with which he seemed well pleased. I
made signs for him to bring us something to eat' and he shortly
returned with some dried salmon, 'which he would give to no
one but me, whom we thought he asked for by name of
Capitane, but in this we were probably mistaken, because I do
not see how he could know that I was the Captain.'

By the middle of September 'it was high time to think of
leaving these northern parts and to retire to some place to spend
the winter where I could procure refreshments for the people
and a small supply of provisions. *Petropaulowska* in *Kamchatka*
did not appear to me a place where I could procure either the
one or the other for so large a number of men, and besides, I
had other reasons for not going there at this time, the first, and
on which all the others depended, was the great dislike I had to
lay inactive for six or seven months, which must have been the

case had I wintered in any of these northern parts.' This was the man who had wintered in Newfoundland so many times in his early naval years. 'No place was so conveniently within our reach where we could expect to meet with these necessary articles as the Sandwich Islands. To these islands, therefore, I intended to proceed.'

Once more the *Resolution* 'sprung a leak which filled the spirit room with water before it was discovered,' so they were obliged to put in to Samgoonoodha and 'the carpenters belonging to both ships were set to work to rip off the sheathing ... where many of the seams were found quite open so that it was no wonder that so much water found its way into the ship.' While this was going on Cook sent 'one third of the people by turns' ashore to pick berries, which were in great abundance, 'so that if there were any seeds of the scurvy in either ship, these berries and spruce beer, which they had to drink every other day, effectually removed it.'

It was now October. 'On the 8th I received by the hand of an Indian named Derramoushk a very singular present considering the place. It was a rye loaf, or rather a pie made in the form of a loaf, for some salmon highly seasoned with pepper was in it. He had the like present for Captain Clerke and a note to each of us written in a language none of us could understand. We, however, had no doubt but this present was from some Russians in our neighbourhood and sent to these our unknown friends by the same hand a few bottles of rum, wine and porter, which we thought would be as acceptable as anything we had besides, and the event proved we were not mistaken.'

A couple of days later the Indian returned 'with three Russian seamen or furriers ... well behaved, intelligent men, and very ready to give me all the information I could desire, but for want of an interpreter we had some difficulty to understand each other. ... One of these men said he was on the America Voyage with Behring, he must, however, have been very young, for he had not now the appearance of an old man. The memory of few men is held in greater esteem than these men do Behring's, probably from his being the occasion of their fur trade being extended to the eastward, which was the consequence of that able navigator's misfortunes, for had not chance and his distresses carried him to the island which bears his name, and where he died, it is probable the Russians would never have thought of making further discoveries on the American coast.'

Another Russian, Gregorioff Sin Ismyloff, came aboard and he said he had been on the expedition led by Sind. 'Where Mr Sind went to after, or how he spent the two years Ismyloff said they were out, he either could not or would not inform us; or else we could not make him comprehend what it was we wanted, and yet in almost every other thing we could make him understand us, which made us suspect that he went a little too far when he said he was in this expedition.' They 'knew nothing of the continent of America to the northward ... they call it by the same name as Mr Sind does his great island, Alaschka.'

Cook visited the Russian fur traders' houses where he found they lived a community life. 'They all live in the same house, the Russians at the upper end, the Kamtschatkdales in the middle and the natives at the lower end, where is fixed a large boiler for boiling their victuals, consisting chiefly of what the sea produceth, with the addition of wild roots and berries. There is little more difference between the first and last table than what is produced by cookery, in which the Russians have the art to make indifferent things palatable. I have eaten whales' flesh of their cooking that I thought very good, and they make a sort of pan-pudding of salmon roe beat up fine and fried, that is no bad succedaneum for bread.'

He found the natives 'lousy and filthy in their houses.' The sleeping places were 'tolerable decent. But the middle of the house, which is common to all, is just the reverse, for although it is covered with dry grass, it is a receptacle for all the dirt in the house and the place for the urine trough, the stench of which is not a bit mended by raw hides or leather being almost continually steeped in it.' He was 'once present when the Chief of Oonalaska dined off the raw head of a large halibut, but just caught. Before any was given to the chief, two of his servants ate the gills without any other dressing than squeezing out the slime ... they then cut large mouthfuls of meat and laid them before the chief, who ate it with the same satisfaction as we should do to a raw oyster.' Cook goes on to note: 'There are few if any that do not both smoke and chew tobacco and take snuff, a luxury that bids fair to keep them always poor.'

'They are remarkably cheerful and friendly amongst each other, and always behaved with great civility to our people. The women grant the last favour without the least scruple; young or old, married or single, I have been told, never hesitate a moment. The Russians told us they never had any connections with the Indian women, because they were not Christians; our people were not so scrupulous, and some were taken in, for the venereal distemper is not unknown to these people.'

As to the North West Passage, Cook concluded 'there can be little doubt but there is a northern communication of some sort by sea between this Ocean and Baffin's Bay, but it may be effectually shut up against shipping by ice and other impediments.' (This is precisely the conclusion which William Scoresby came to forty years later*.)

It was now the end of October and the day before his fiftieth birthday 'in the morning of Monday the 26th we put to sea. My intention was now to proceed to the Sandwich Islands to spend a few of the winter months, provided we met with the necessary refreshments there, and then proceed to Kamtschatka, endeavouring to be there by the middle of May next.'

*Cf; our book *William Scoresby Arctic Scientist* p 52

CHAPTER 14

THE END OF A LIFE

In every situation he stood unrivalled and alone;
on him all eyes were turned; he was our leading star,
which at its setting left us involved in darkness and
despair.

Samwell.

All this time the faithful support ship the *Discovery* with
Captain Clerke, accompanied the *Resolution*. We have said little
about it in this book since it is primarily intended to be a
portrait of James Cook the man and to show how he revealed
himself in his writing. Perhaps we have tended to take the
Discovery for granted, so quietly and efficiently did her captain
follow his great leader. Charles Clerke's work is even more
meritorious when we know that he was a sick man, doomed to
die of tuberculosis, very probably contracted in the cabin of
Hicks on the first voyage.

On the southward passage they ran into severe gales and
'before night blew a violent storm which obliged us to bring to.
The *Discovery* fired several guns, which we answered, but without
knowing on what occasion they were fired.' The following day,
'there being but little wind, Captain Clerke came on board and
informed me of a melancholy accident that happened on board
his ship . . . the main tack gave way, killed one man outright
and wounded the Boatswain and two or three more . . . his
sails and rigging received considerable damage and that the guns
which he fired was the signal to bring to.'

Before long Cook himself experienced similar trouble. 'What
made our situation more alarming was the leach rope of the
main topsail giving way, which was the occasion of the sail
being rent in two and the two topgallant sails gave way in the
same manner though not half worn.'

The following passage was not included when the earlier
versions of Cook's Journal were published: 'On this occasion I
cannot help observing that I have always found that the bolt
ropes of our sails have not been of sufficient strength or

substance to even half wear out the canvas: this at different times has been the occasion of much expense of canvas and infinite trouble and vexation. Nor are the cordage and canvas, or indeed hardly any other stores made use of in the Navy, of equal goodness with those in general use in the Merchant service. ... These evils are likely never to be redressed for, besides the difficulty of procuring stores for the Crown of equal goodness with those purchased by private people for their own use, it is a generally received opinion amongst naval officers of all ranks that no stores are equal in goodness to those of the Crown ... but it is in the quantity and not in the quality of the stores, this last is seldom tried, for things are generally condemned or converted to some other use by such time as they are half worn out. It is only on such voyages as these we have an opportunity to make the trial where everything is obliged to be worn to the very utmost.'

It is interesting to note that when the manuscript of the Journal containing this criticism of Naval stores came to Sir Hugh Palliser's notice, his loyalty to the service took precedence over his friendship and knowledge of Cook's veracity. His suggestion to omit the passage from the publication was accepted for 'it is well known,' he wrote, 'that there is no better cordage than what is made in the King's Yards.'

We see in this how far Cook was detached from conventional naval ideas and attitudes and how the urgency of practical considerations leading to efficiency always came first in his mind. Had Palliser or other influential naval persons acted on this clear criticism of Cook's, much might have been accomplished in improving the naval supply system and ridding it of the corrupt practices then widely prevalent.

It was about this time that another casualty occurred. A favourite black cat fell overboard, but as they were near land at the time some natives rescued it and brought it to the ship in their canoe, for which they were suitably rewarded.

They reached the Hawaiian islands in December. Here Cook 'procured a quantity of sugar cane and found that a strong decoction of it made a very palatable beer ... but not one of my mutinous crew would even so much as taste it. As I had no motive for doing it but to save our spirit for a colder climate, I gave myself no trouble either to oblige or persuade them to drink it, knowing there was no danger of the scurvy so long as we had plenty of other vegetables; but that I might not be disappointed in my views, I gave orders that no grog should be served in either ship. Myself and the officers continued to make use of this beer ... though my turbulent crew alleged it was injurious to their health.'

'Every innovation whatever, though ever so much to their advantage, is sure to meet with the highest disapprobation from seamen. Portable soup and sour krout were at first condemned by them as stuff not fit for human beings to eat. Few men have introduced into their ships more novelties by way of victuals and drink than I have done, indeed, few men have had the same opportunity or been driven to the same necessity. It has, however, in a great measure been owing to such little innovations that I have always kept my people generally speaking free from that dreadful distemper the scurvy.'

The New Year 'was ushered in with very hard rain, which continued at intervals till past ten o'clock. Being at this time about five miles from the land, several canoes came off with fruit and roots and at last with hogs, though not many.' As they sailed around Hawaii, charting the land, this was the general pattern. Later, 'as it was a fine, pleasant day we had plenty of company and abundance of everything. We had the company of several all night and their canoes towing astern.' A couple of nights later: 'as the night approached the Indians retired to the shore, a good many, however, desired to sleep on board. Curiosity was not their only motive, at least not with some of them, for the next morning several things were missing, which determined me not to entertain so many another night.'

The final entry in Cook's *Journal* was made on 17th January 1779, when they were anchored in Kealakekua Bay. Both the ships were 'very much crowded with Indians and surrounded with a multitude of canoes. I have nowhere in this sea seen such a number of people assembled at one place. Besides those in the canoes all the shore of the bay was covered with people and hundreds were swimming about the ships like shoals of fish.'

The thieving habit of the natives was the indirect cause of Captain Cook's death, for when they stole one of the ship's boats he was absolutely determined to recover it. On the 14th of February, 1779 he went ashore to demand its return, and amongst a vast crowd of savages, as he turned away for a moment, he was struck, fell, and was clubbed to death.

His end has been too often told not to be known. From the

many accounts of those present it is abundantly clear that he was brutally hacked to death by a mob of savages, who then dismembered his body. With difficulty, parts of it — the hand with the well known scar among them — were brought back to the *Resolution* and they were 'committed to the deep with all the attention and honour we could possibly pay it in this part of the world.'

This is from the *Journal* of Captain Charles Clerke, who succeeded Cook. 'I cannot help,' he wrote, 'lamenting my own unhappy state of health which sometimes is so bad as hardly to suffer me to keep the deck and of course farther incapacitates me for the succeeding so able a Navigator as my honoured friend and predecessor.' Although Clerke had 'some notion of taking a stout party ashore, make what destruction among them I could, then burn the town, canoes &c' he was prevailed upon not to do this.

The account given by King tells us: 'Our opinions were very different; unfortunately he that was to determine, although with the best intentions, was in so feeble a state of health as to be unable to act from his own ideas only, and he listened too much to those of others. Whilst one party was for carrying the ships close to the town and making all preparations for a vigorous chastisement, others were for mild measures till the mast was repaired & the ships in a condition for sailing and then to take an ample revenge; whilst a third was for endeavouring by all means to become & to continue friends, as the injury was already done and irreparable.'

As can well be imagined, the crews of the ships thirsted for revenge, and they set fire to and completely destroyed an entire village before they could be restrained.

At last the repairs were finished and on the 23rd February 1779 they left Kealakekua Bay, 'a place become too remarkably famous for the very unfortunate & tragical death of one of the greatest Navigators our Nation, or any Nation, ever had.' (Clerke)

Samwell, the surgeon's mate, wrote a comprehensive and moving testimony to Cook in which he said: 'He was beloved by his people, who looked up to him as a father and obeyed his commands with alacrity. ... In every situation he stood unrivalled and alone; on him all eyes were turned; he was our leading star, which at its setting left us involved in darkness and despair.'

It was almost a year before the news reached England, by
way of letters from Russia. The dying Clerke, faithful to Cook's
instructions, took the ships back to the Arctic to search once
more for an opening in the ice. They called at Petropavlovsk
and found the inhabitants to be suffering acutely from scurvy,
following a severe winter without fresh vegetables. Captain
Clerke ordered the sick people to be put under the care of the
surgeons of the ships who, according to Young, 'treated them
with sour krout, sweet wort &c, according to the method of
Captain Cook and their recovery was speedy and surprising.'

Captain Charles Clerke ended his thirty-eight years in August
1779, in sight of Kamchatka, where he was buried ashore.

After his death the command went to the unpopular
American John Gore, who brought the ships back by way of
Japan, Sumatra and the Cape. Unable to get into the Channel,
he sailed up the west coast of Britain and stopped at the
Orkney Islands; 'but the reason of our putting in there,' wrote
Gilbert, 'when we had water enough on board and a favourable
wind to carry us round to the River, was known only to our
Commander.'

Contrary winds kept them in Stromness a further month,
during which time the Sergeant of Marines, Samuel Gibson, —
veteran of three voyages — saw fit to get married. Alas, he
died on the southward passage to London, where they arrived
on the 4th October 1780. Gilbert's words conclude the third
volume of the Journals, the life work of J. C. Beaglehole:

'Thus ended a long, tedious and disagreeable voyage of four
years and three months, during which we lost only seven
persons by sickness and three by accident . . . exclusive of those
that were killed with our great and unfortunate Commander.'

At the very same time that the ships were limping home to
the Thames there was a fearful hurricane off the coast of
Jamaica in which thirteen ships of the Royal Navy were lost.
One of those which was sunk with all hands was the *Thunderer,*
(built at Deptford, not far from the Cook home) and serving in
her was Nathaniel, the second son of James and Elizabeth Cook,
He was fifteen years old.

Mrs Cook was now left with only her two sons, the baby,
Hugh, and her firstborn, James. He made a very promising
career in the Navy and reached the rank of commander by the
time he was thirty. Hugh was destined to go into the Church
and he entered Christ's College, Cambridge. There, he died of

a fever a few days before Christmas 1793. He was seventeen years old.

Only a month after, James lost his life on the 25th January 1794 in going from Poole to the *Spitfire* sloop of war, which he commanded. He was thirty years old.

The effect of all these blows upon Mrs Cook can only be imagined. Bearing her grief, keeping fast-days on all the anniversaries, she lived on well into her nineties. For the greater part of her old age, her cousin, Isaac Smith, who became an admiral, resided with her. She died on 13th May 1835.

CONCLUSION

On his return from his third voyage Cook's achievements would have lifted him still higher and his prestige and standing would have been such that any project thought worthy of his attention would have had the support and enthusiastic backing of his countrymen from the Crown down to the humblest seaman. But it was not to be. There was no return. Only the memory of past achievements, great and unparalleled, and a sense of irreparable loss.

Theories have been put forward, regarding Cook's death, attempting to relate it to the religious and political circumstances of the island people and speaking of a struggle for power between a declining elderly priesthood and a vigorous younger faction of chiefs. There has been a supposed divinity of Cook and a possible reincarnation. Some wild theorists have even supposed he was killed by his own men. It is difficult to see. two centuries later, how we may arrive at any certain knowledge of these matters. What does seem indisputable is that Cook was murdered by a mob of savages; excited primitive people driven to excesses by extremists, — circumstances not unknown to later times and having always the same fatal irreversible consequences.

'Everyone in the ship,' says Zimmerman*, 'was stricken dumb, crushed and felt as though he had lost his father.' He tells us that all the crew shed tears when they committed the fragments of his body to the sea. But the disposal of the remains, however distressing for those involved, is now only an incident in history — a brief and awful interlude. The life that had been lived was one of peaceful achievement, of penetration into the unknown, of placing of knowledge and certainty where there had been ignorance and doubt, and of an ordered and disciplined life in the service of humanity. This is what matters in the end, the quality of the life that is given to each one of us.

When the awful news eventually reached Mrs Cook it must have been more than distressing. She was little different from many other sailors' widows in having no grave over which she could weep; but the thought of his tenderly loved body falling

---

*Zimmerman was a German seaman who went the last voyage and wrote a journal which he smuggled out and later published in German. It was translated only comparatively recently.

into cannibals' hands must have seared her through and through. This splendid man whom she had known better than he had been known by any other human being, who had shared her life in a few, all-too-brief, never to be forgotten times, was no more. And she had to live on another fifty-six years, bereft of her entire family for the greater part of them.

The sudden end of the one fills us with horror. The long and lonely vigil of the other, with pity and compassion. Let us not dwell further on it, but rather on the man's achievement.

Cook's life falls into four distinct parts. Firstly, his formative years in Cleveland and Staithes; next his coming to Whitby and his apprenticeship to the sea; thirdly, his joining the navy and the Quebec and Newfoundland period, during which his growing skills, acquired knowledge and real experience fitted him particularly for the fourth and final part, — his great voyages.

This, of course, is to simplify, but how can one write about any man's life without a certain degree of simplification? Life is so complex, meetings are so numerous, influences so pervasive and experiences so many and varied that the kaleidoscopic patterns of a man's life present many different combinations. The ever recurring pattern of Cook's life is of a man completely in control of himself and his situation, facing all contingencies with the lively assurance that a way will be found round all difficulties through a ready application of knowledge and experience, for we see in Cook the great value of a balanced and well informed optimism.

The child is father of the man and early environments give an unalterable shape to the life of a man. The great man that was James Cook was born among the hills of Cleveland, nurtured in the countryside and farm life of Aireyholme; glimpsed the world of knowledge in the schoolroom at Great Ayton; caught his first sight of the sea from Staithes; was won to it there among the fishing folk and fishing craft and remained captive to it ever after. From Whitby he sailed and matured into a man of the sea. Here, in the Grape Lane house he studied, and shared the Quaker life and times of his master and friend. He absorbed the life of the old seaport into his very fibres, — its work and its ways; its ship building and repairing; its sea and river traffic; its ship chandling and rope making and all its sea life and sea ventures of the times of his early youth and manhood.

That Cook was a humane and civilised man owes much, as

indeed it always does, to innate qualities of mind and heart, but the influence on him during the formative years of the Whitby Quaker community and especially the Walker family, was immensely important. In the accepted forms, Zimmerman tells us, he was not religious. 'He never spoke of religion, would tolerate no priest on his ship, seldom observed the sabbath, but otherwise was a just man in all his dealings; he never swore, not even when in the greatest anger. ... Temperance was one of his chief virtues. ... His table was sparsely laid, much more so than that of his officers.'

This could have been said of any Quaker. His habit of silent inner reflection ('He would often sit at the table without saying a word and was always very reserved') was a typical Quakerly attribute.

In these periods of quiet thought Cook gained those insights and formulated those plans which, later translated into action, so often astonished his companions. The masterly professional competence of Cook, while being a natural consequence of his intellect, training and wide experience also owed much to this deep inner reflectiveness. That almost uncanny knowledge of what lay ahead was Cook's response to the many subtle impressions and influences of which he was aware but which went unheeded or uninterpreted by all around him. He alone seemed to have an intuitive awareness of what lay just beyond the horizon.

It has of course been recognised from his own times onwards that Cook was one of the greatest navigators and explorers of all time. How does the son of an obscure farmer in the England of the early eighteenth century eventually blaze a trail across the world like a splendid maritime meteor?

The times and the man came together, and the greatness of the one matched the growing, vigorous expanding world of the other. What is the nature of Cook's greatness? We have called him a maritime scientist because we believe the magnitude and nature of his achievements are of the same order as those of a first rate scientist.

Benjamin Franklin fully appreciated the scientific importance of Cook's work, for at the time of the revolutionary war he arranged that his ships should not be molested by American privateers. This act of generous and enlightened statesmanship gives us the measure of both Franklin and Cook. But fine minds have always been in accord with one another and the great American statesman and scientist was well fitted to appreciate James Cook.

In his approach to the vital, practical problem of preserving the health of seamen on long sea voyages; in his competence as a working astronomer and in his mastery of the lunar method of finding longitude at sea; in his rigorous testing of the Harrison-Kendall chronometer; but more especially in his unrivalled skills as navigator, explorer and cartographer, Captain Cook epitomises the wide ranging curiosity and inspired common sense of the best minds of his time. Men like Herschel and Priestley, who were at the apex of the growing, advancing spirit of their age and sought with all the resources at their command to further the finest aspirations of their times. The measure of their success is a measure of the vigour and vitality of the nation to which they belonged. This was the time when men with brains were drawn to Britain and Herschel was only an outstanding example among the many who came.

There are many similarities in the lives of the two men, Herschel and Cook. Both made outstanding efforts of self-help in education and mastered the manipulative and technical skills appropriate to their own particular fields of endeavour. By becoming more and more competent and professional, they began to make contributions to the science of their time.

There is no doubt, and indeed it has become increasingly evident over the years, that Cook largely fulfilled one of the major aims of the science of his time, namely, the discovery and delineation of the main outstanding features of the world of sea and land masses. This comprised his discovery of the long eastern shoreline of Australia, accomplished by his altogether breathtaking navigation in the Great Barrier Reef; his astonishing outline maps of the two islands of New Zealand; his examination of long stretches of the coastline of North America; the discovery of completely new islands, such as Hawaii and New Caledonia, and the finding and accurate placing of island groups previously known in vague or indeterminate fashion.

Cook's penetration into far southern waters, his continuous and persistent efforts to determine the nature and extent of Antarctica, make him the first navigator to set close limits to that continent and even if he did not actually see any of it, he established that it lay under eternal ice and snow, was inhospitable, remote and savage almost beyond all human reckoning.

So too in northern waters, Cook sailed into Bering Strait,

confirmed its nature and characteristics, and showed the great practical difficulties of navigation in these regions. No one before Cook had penetrated so far in time and space in long sea voyaging, nor brought back so much detailed and accurate knowledge. Clearly this could only have been done by a man of outstanding character and genius. To accomplish all this, he changed the nature and form of long sea voyaging for ever, both in regard to the accurate determination of position and – most vital – the health of crews.

Despite Lind's excellent treatise and Bachstrom's masterly paper, there was still much doubt and uncertainty in the medical mind; there were many pet theories and obscure doctrines. Those having the care and control of sea diet were offered much conflicting and confusing advice. Cook cast all the dubious rigmarole aside and concentrated on cleanliness and sour krout and fresh food whenever available. In the second voyage he vindicated these cardinal controls completely. The Fellows of the Royal Society recognised the magnitude of this breakthrough and their award of the Copley Medal to Cook was richly deserved and it reflects credit on their sagacity and insight. It is significant that the President of the Royal Society at this time was a distinguished medical man Sir John Pringle.

We have asked the question, How did Cook achieve greatness? Here we are brought face to face with the miracle of human accomplishment. The really outstanding men are always at one with the spirit of their age and combine within themselves all the major qualities necessary for success in their chosen field. If we follow Cook's career from its humble beginnings, we see how step by step he won for himself all that was needful, – by careful and persistent study; by active and vigorous work; by eager willingness to participate in any field of endeavour where positive and worthwhile results might be obtained; by his joining the Navy through an act of patriotism and a love of adventure, which eventually gave him that professional standing and competence which opened the door to great achievements, – and always he maintained a courageous fortitude in times of difficulty and stress.

Sometimes a man's gifts are singularly fitted to the needs and opportunities of his times. Cook's were of this nature. This is not to diminish in any way the magnitude of what he did. To be made by and to make the times in which one lives is the measure of greatness; and when the life is one of peaceful

CAPTAIN COOK'S MONUMENT. Given to the town of Whitby by Gervase Beckett, M.P., the statue, the work of John Tweed, stands on the West Cliff. Cook, a map in one hand and dividers in the other, looks, — as the sculptor intended — out to sea, his eyes on the distant horizon.

achievement, an enlarging of the horizons of human knowledge and experience, then the greatness has a permanence which outlives time and circumstance and becomes a part of the history of the real progress of humanity. It is not a catalogue of battles or political victories, but a record of real advance.

After such a life and work the world picture is changed and humanity may move forward to fresh conquests, secure in what has been accomplished, and given fresh encouragement in the life that has been lived.

Cook's life span comes between the death of Newton and the birth of Darwin. Our country is fortunate that it may count such names among its sons.

CHRONOLOGY

1728	27 October. Born at Marton-in-Cleveland.
1735	Family move to Aireyholme Farm.
1740	School at Great Ayton.
1745	In Sanderson's shop at Staithes.
1747	Whitby. Apprenticed to John Walker.
1752	Mate of the *Friendship*.
1755	Joins the Navy. HMS *Eagle*.
1757	27 October. Master of HMS *Pembroke*.
1759	Quebec. Newfoundland.
1762	21 December. Marries Elizabeth Batts.
1763	Surveyor of Newfoundland. HMS *Grenville*. October. First son, James, born. (Drowned 1794).
1764	December. Second son, Nathaniel, born. (Drowned 1780).
1766	Eclipse observed at Cape Race.
1767	Daughter, Elizabeth, born. (Died 1771).
1768	August. Third son, Joseph, born. (Died Sept. 1768).

FIRST VOYAGE

1768	26 August. *Endeavour* sails from Plymouth.
1769	June. Transit of Venus observed at Tahiti. October. New Zealand.
1770	Botany Bay. Great Barrier Reef.
1771	*Endeavour* returns to England.
1772	June. Fourth son, George, born. (Died October 1772).

SECOND VOYAGE

1772	13 July *Resolution* and *Adventure* sail.
1773	Antarctic circle crossed.
1774	*Adventure* returns home. Easter Island. Norfolk Island. New Zealand. Circumnavigation of south seas.
1775	30 July *Resolution* returns to England.
1776	Elected F.R.S. May. Fifth son, Hugh, born. (Died 1793).

THIRD VOYAGE

1776 12 July. *Resolution* and *Discovery* sail.
1777 New Zealand. Tonga. Tahiti. Christmas Island.
1778 Hawaii. Alaska. Nootka Sound. Bering Strait.
1779 14 February. Murdered at Hawaii.

1780 4 October. Remnants of the expedition arrive at Plymouth.
1835 Death of Mrs Elizabeth Cook.

APPENDIX I

BACHSTROM JOHANN FRIEDRICH 1686–1742

Lutheran theologist, doctor of medicine, inventor, writer humanist and educationalist of the school of Comenius, he was also a student of natural history and physics.

He was born in Rawicz in the kingdom of Poland and was the son of a prominent goldsmith.

Educated at the Elizabeth Grammar School, Breslau; 1710 studied in Jena under the Comenius scholar, J. F. Buddeus; Graduated in Copenhagen with a thesis on the Polish Plague.

Bachstrom wandered all over Europe seeking tolerance and freedom from religious bigotry. He wished his daughters to study medicine but failed to get them admitted to the medical schools of either Halle or Leipzig. He was obliged to flee to Constantinople and there he was active in spreading Christian ideals, working in the Turkish educational system. He tried to set up a printing press, but here again conditions became impossible and he resolved to go to a strictly protestant area.

In Breslau, Gorlitz, Freiburg, Dresden and later in Holland and England, he undertook studies in theology, physics and medicine. His natural history studies led to a correspondence with J. Fr. Henkel of Freiburg.

During these years he invented a swimming apparatus to help shipwrecked sailors. Out of his experiences in trying to educate his daughters he elaborated advanced views on the education of women and their place in society.

He wrote a Utopia in the accepted literary form of the period.

In 1737 he went to Lithuania, where he was befriended by the aristocratic Radizwill family, whose son, Heronymus he cured of a speech defect through a self-devised therapy. They made him director of their porcelain and mirror factory. He was the confidant and friend of the duchess Anna, but the son Heronymus turned on his benefactor and, urged on by the Jesuits, accused him of high treason. Bachstrom was thrown into prison and one June morning in 1742 he was found dead in the dungeon.

From a TREATISE OF THE SCURVY by James Lind M.D.
FRCP(E) Published in 1753. Page 314

1734 Observationes circa scorbutum: ejusque indolem, signa
et curam. Auctore, Johann Friedrich Bachstrom.

From want of proper attention to the history of the scurvy,
its causes have been generally, though wrongly, supposed to
be: cold in northern climates, sea air, the use of salt meats &c:
whereas this evil is solely owing to a total abstinence from
fresh vegetable food and greens: which is alone the true primary
cause of the disease. And where persons, either through neglect
or necessity, do refrain for a considerable time from eating the
fresh fruits of the earth, and greens, no age, no climate or soil
are exempted from its attack. Other secondary causes may
likewise concur, but recent vegetables are found alone effectual
to preserve the body from this malady; and most speedily to
cure it, even in a few days, when the case is not rendered
desperate by the patient's being dropsical or consumptive. All
of which is founded on the following observations.

He remarks that the scurvy is most frequent among northern
nations and in the coldest countries. There it is not confined to
the sea alone, but rages with great violence at land, affecting
both natives and foreigners, of which the poor seamen left to
winter in Greenland, who were all cut off by this distemper,
afford a memorable instance. But the opinion of its being
produced there by cold, he thinks irreconcilable with the
daily experience of its attacking seamen in their voyages to
the Indies, even when under the torrid zone.

That it is not peculiar to the sea the following histories
sufficiently evince. During the late siege of Thorn 5 or 6000
died . . . of this dreadful calamity. No sooner was the siege over
and the gates of the town open for the admission of vegetables
and greens from the country, but the mortality quickly ceased
and the disease at once disappeared.

In the end of the last war with the Turks . . . many thousands
of the common soldiers (but not one officer, as having different
diet) were cut off by the scurvy. It persisted until the spring,
when the earth was covered with greens and vegetables.

A sailor in the Greenland ships was so over run and disabled with the scurvy that his companions put him in a boat and sent him ashore. The poor wretch had lost the use of his limbs and could only crawl about on the ground. This he found covered with a plant which he, continually grazing like a beast of the field, plucked up with his teeth. In a short time he was by this means perfectly recovered; and upon his return home, it was found to have been the herb, scurvy grass.

From all which, the author concludes, that as abstinence from recent vegetables is altogether and solely the cause of the distemper, so these alone are its effectual remedies. Accordingly, he bestowes the epithet of antiscorbutic on all that class which are wholesome and eatable; observing Nature everywhere affords a supply of remedies, even in Greenland and the most frozen countries. There no sooner the snow melts from the rivers, but their borders are covered with brooklime, cresses and scurvy grass in ample prodigality.

For prevention he recommends living much upon green vegetables when they can be got: otherwise upon preserved fruits, herbs, roots etc. He advises seamen when at land to be more careful of laying up a store of greens than of flesh; and in case of necessity would have them, when at sea, to make trial of the seaweeds that grow upon the ship's bottom, being persuaded that the great physician of nature had not left them without a remedy, although he had never heard of its being tried.

He condemns the use of steel, mercury and alum, as likewise sulphurous and vitriolic medicines, especially the strong acid of vitriol, which some account a specific in the scurvy; but they will find themselves disappointed.

The most common herbs and fresh fruits excel the most pompous pharmaceutical preparations, especially those of the animal and mineral kind.

APPENDIX III

QUAKER ADVICES
From earliest times Advices were sent to small Quaker meetings; later they were supplemented by Queries (Do you cherish that of God within you? — Do you cherish a forgiving spirit? — Are you honest and truthful in word and deed?) as well.

Neither were intended to be rigid rules but, in the words of one of the earliest of these Advices, sent to 'The brethren in the North' in 1656; 'Dearly beloved Friends, these things we do not lay upon you as a rule or form to walk by; but that all with a measure of the light, which is pure and holy, may be guided; and so in the light walking and abiding, these things may be fulfilled in the Spirit, not in the letter, for the letter killeth, but the Spirit giveth life.'

The following are just a few extracted from the many which were issued, and they are contemporary with Cook's years with the Walkers.

YOUTH We earnestly beseech our friends, and especially the youth among us, to avoid all such conversation as may tend to draw out their minds into the foolish and wicked pastimes with which this age aboundeth (particularly balls, gaming places, horse races and play houses) those nurseries of debauchery and wickedness, the burthen and grief of the sober part of other societies, as well as of our own; practices wholly unbecoming a people under the Christian profession.

PLAINNESS Take care to keep to truth and plainness, in language, habit, deportment and behaviour. . . . Avoid pride and immodesty in apparel, and all vain and superfluous fashions of the world.

Our testimony is against an undue liberty . . . by the extravagant head dress and apparel of many of both sexes. And in like vanity of mind, divers amongst us run into great extravagancies in the furniture of their houses. . . . Keep to that which is modest, decent, plain and useful.

OATHS Advised that our Christian testimony be faithfully maintained against the burthen and imposition of oaths, according to the express prohibition of Christ and also of the apostle James:

'But I say unto you, swear not at all; neither by Heaven, for it is God's throne; nor by the earth for it is his footstool ... But let your communication be, Yea, yea; Nay, nay.' *Matt. v 33–37*
'But above all things my brethren, swear not; neither by heaven, neither by the earth, neither by any other oath: but let your yea be yea; and your nay, nay.' *James v 12.*

MASTERS, MISTRESSES AND SERVANTS. A religious care is recommended toward our servants, that all appearance of pride, idleness, and vain conversation in them may be discouraged, and that they may be exhorted to attend first day meetings, and have a sense of God's love upon their spirits, and therein partake with us of the sweetness of truth.

CIVIL GOVERNMENT. Walk wisely and circumspectly towards all men, in the peaceable spirit of Christ Jesus, giving no offence or occasions to those in outward government, nor way to any controversies, heats and distractions of this world, about the kingdoms of it. It is advised that friends be circumspect and not make it their business to discourse of the outward powers, but to discourage all such things.

CONDUCT AND CONVERSATION. Be prudent in all manner of behaviour, both in public and private; avoiding all intemperance in eating and drinking, and likewise foolish jesting. Let our moderation and prudence, as well as truth and justice, appear to all men, and in all things, in trading and commerce, in speech and communication, in eating and drinking, in habit and furniture, and through all in a meek, lowly, quiet spirit.
Frequent waiting in stillness on the Lord for the renewal of strength, keeps the mind at home in its proper place and duty, and out of all unprofitable association and converse. Much hurt may accrue to the religious mind by long and frequent conversation on temporal matters, especially by interesting ourselves too much in them; for there is a leaven therein, which, being suffered to prevail, indisposes and benumbs the soul, and prevents its frequent ascendings in living aspirations towards the fountain of eternal life.
Let your behaviour among men be unblamable. Let not the vain and foolish fashions and customs of the world prevail over you. Avoid sports, plays and all such diversions, as tending to

alienate the mind from God, and to deprive the soul of his comfortable presence and power. Be temperate and sober; shun all excess in eating and drinking; and let your moderation be known to all men.

BIBLIOGRAPHY

The Life and Voyages of Captain James Cook
by the Rev. George Young   London 1836

The Life of Captain James Cook
J. C. Beaglehole   London 1974

The Journals of Captain James Cook.   4 vols.
Edited by J. C. Beaglehole and R. A. Skelton   Cambridge 1969

Captain Cook, the Seaman's Seaman
Alan Villiers   London 1967

Cook and the Opening of the Pacific
James A. Williamson   London 1946

Sir Joseph Banks, The Autocrat of the Philosophers
H. C. Cameron   London 1952

Scientists of the Industrial Revolution
J. G. Crowther   London 1962

Omai, Noble Savage
M. Alexander   London 1977

A History of Whitby (1817)
George Young   reprint Whitby 1976

Zimmermann's Captain Cook
Edited by F. W. Howay   Canada 1930

Captain Cook's Voyages of Discovery
John Barrow. Everyman edition.   London 1860

Captain Cook
Sir Walter Besant   London 1890

Captain James Cook
Arthur Kitson   London 1907

A Treatise of the Scurvy
James Lind F.R.C.P.(E). 1753   reprint Edinburgh 1953

History and Antiquities of Cleveland
J. W. Ord   London 1846

Note: The above titles are books which we have read. For a comprehensive list of further reading see the bibliography given by J. C. Beaglehole in his *Life of Captain James Cook.*

INDEX